LEARNED AND APPLIED

This book is a tribute to Prof. Dr. Arnold Verruijt, at the occasion of his retirement as professor in soil mechanics at the Delft University of Technology on 25 january 2002

LEARNED AND APPLIED

SOIL MECHANICS OUT OF DELFT

EDITED BY

FRANS B.J. BARENDS & PAOLA M.P.C. STEIJGER

Delft University of Technology, Delft, The Netherlands

A.A. BALKEMA PUBLISHERS / LISSE / ABINGDON / EXTON (PA) / TOKYO

ISBN 90 5808 357 3

Contents

Geodynamics

Geostatics

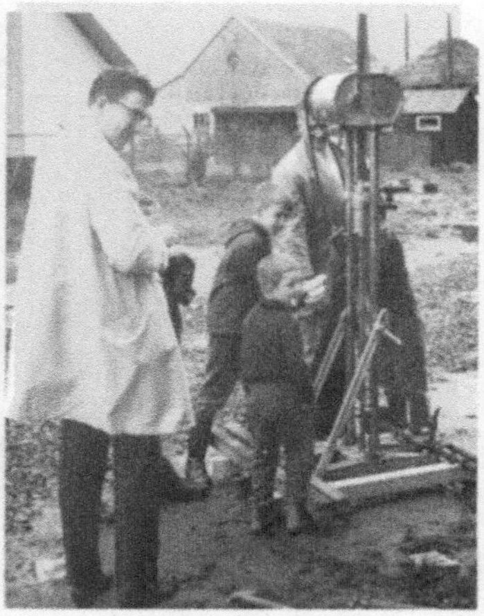

1967

1971

1982

1983

1983

Soil Mechanics out of Delft

Tribute to Professor Dr Arnold Verruijt

FRANS B.J. BARENDS
TUDelft & GeoDelft, Delft, the Netherlands

This book is a result of a joined effort in the occasion of the retirement of Arnold Verruijt as professor at the Technical University Delft by the group of promovendi who completed their doctorate under his guidance in the past 30 years. It is divided in 5 chapters, areas of special interest of Arnold Verruijt: groundwater flow, consolidation, numerical methods, geodynamics and geostatics. Dr Arnold Verruijt is professor of soil mechanics since 1975, a beloved teacher, the first to received the Delft University Master Prize in 1994, and author of more than 100 technical papers and several educational books, which show all remarkable brightness and style. They are a pleasure to read.

In this tribute a summary is presented on Verruijt's scientific contributions. Amongst unprecedented achievements, the inclusion of pore-fluid compressibility in the theory of consolidation, the flushing effect in interface flow, the Noordbergum effect in groundwater recovery, the concept of hysteretic damping in geodynamics, and analytical solutions of elastic deformations in tunneling are to be mentioned. He was one of the firsts to develop practical geotechnical computer programs. At present, he is a pioneer in using internet for education. We know Arnold as a friendly, intelligent and eloquent colleague, our national torch-bearer in the international field of geotechnics, a prominent member of the Royal Dutch Academy of Sciences, and Officer in the Royal Order of Oranje-Nassau.

Review of the scientific work of Dr Arnold Verruijt

Groundwater flow

Salient is Verruijt's textbook *Groundwater Flow* [1], originally written in 1970 for courses on the fundamentals of the theory of groundwater flow. In a second edition in 1982 unsteady flow and numerical methods are introduced. Humbly, the introduction of the book states that most of the material presented was not developed by him. However, the value of the book lays in the way it is written: concise, comprehensive, clear, and self-explanatory.

His interest in groundwater focuses on fundamentals of flow and transport, phreatic and interface problems, and solution methods, particularly conformal mapping and finite element techniques [2]. He discovered a peculiar phenomenon in interface flow: the flushing effect [3]. In short. The width of the dispersion zone across an interface separating two fluids in a porous medium is reduced by the relative velocity of both. When a fresh fluid ($c=0$ at $x=0$) flows with velocity v along an interface at $y=0$, $x>0$ above stagnant salty water ($c=c_0$ for $x>0$, $y<0$), the dispersion can be described by

$$c = c_0\mathrm{erfc}[\xi] - \eta\exp[-\xi^2]; \quad \xi = y/\sqrt{[4x\lambda]}; \quad \eta = \alpha v(\lambda+2\mu)/\sqrt{[\pi x/\lambda]}$$

where α is the transfer resistance coefficient at the interface and λ and μ are the transversal and longitudinal dispersion coefficient in the fresh fluid. The second term $\eta\exp[-\xi^2]$ represents the flushing effect. As it appears now, this phenomenon has distinct implications on the long-term behaviour of groundwater quality in the Dutch coastal zone.

Consolidation

One of the methods to solve analytically the three-dimensional linearized consolidation equations is Biot's method, which was developed for three-dimensional thermo-eleasticity by the introduction of a scalar function ϕ and a vector function ψ. Verruijt proved in 1969 [4] the theoretical completeness of Biot's theorem and he showed in 1971 that it can be used also for compressible pore fluids. In fact, according to the theorem, if the field is described by displacement and stress

$$u_i = (\phi + x_j\psi_j)_{,i} - \omega\psi_i \; ; \qquad \sigma_{ij} = -(\lambda+2\mu)(\phi_{,ii}+2\psi_{i,j}) + 2\omega(\lambda+\mu)\psi_{i,j}$$

with $\omega = (\lambda+2\mu)/(\lambda+\mu)$, then the equilibrium conditions imply that the functions ϕ and ψ satisfy

$$\psi_{i,jj} = 0; \; \phi_{,iit} = c\phi_{,iijj} \qquad \text{with} \qquad c = k(\lambda+2\mu)/\gamma$$

where k is the permeability, γ the specific water weight, and λ and μ are Lamé's constants. Verruijt [5] proved that the same holds for a pore fluid with a compressibility β by the introduction of slightly different factors

$$\omega = (1+n\beta(\lambda+2\mu))/(1+n\beta(\lambda+\mu)) \quad \text{and} \quad c = k(\lambda+2\mu)/(\gamma(1+ n\beta(\lambda+2\mu))$$

Solutions can be obtained by integral transformation techniques (Fourier, Hankle, Laplace). Such methods were popular thirty years ago, in the pre-computer era. Verruijt is an expert in such techniques. He applied it to explain a peculiar problem noticed by the Dutch Institute for Drinking Water in the village Noordbergum, where pore pressures surprizingly increased after starting a pumping well [6]. This effect shows similarity with the Mandle effect related to squeeze, and the Cryer effect related to spherical compression. We should call the Noordbergum effect, accordingly, the Verruijt effect.

In a comprehensive article *The theory of Consolidation* Verruijt [7] describes the linearized theory including also the compressibility of individual grains, and, doing so, he sheds new light on the effective stress concept in classical soil mechanics. It complies with the approach in reservoir engineering and has implications on induced landsubsidence [8]. A convenient and practical engineering formula is presented to estimate the rate of pore pressure dissipation for general three-dimensional consolidation problems in a simple manner, just by spatial integration [9].

Numerical Methods

In 1969 Verruijt was visiting professor of hydrology at the University of Toronto, Ontario, for one year. Since this visit a continuous stream of computer models and discussions on their background and applications started. First, in Fortran on main frame computers until 1980, in Basic for microcomputers until 1987, and since then for personal computers in Pascal and via Internet. Verruijt is an expert in finite element techniques, reviewing solution procedures, numerical stability and accuracy [10]. His recent book *Computational Geomechanics* [11] reviews numerical methods for solving a wide range of geotechnical problems, such as flow and transport, elasticity, plasticity and consolidation, for sheetpiling, piles, walls and beams in relation with soil.

Geodynamics

The description of cyclic hysteretic damping in terms of a viscous element by the definition of a retardation factor t_r inversely proportional to frequency ω is a remarkable finding of Verruijt [12]. By comparison of the plastic work done by sliding in a hypothetical volume of four equal spherical particles, cyclically moving from left-trapezoidal through cubic to right-trapezoidal structures, and vice versa, a relation is found for t_r and the damping ratio D, a material constant for soft soil varying between 0.1 and 0.2. As such, introduced in a general visco-elastic constitutive law, $\sigma = G(\varepsilon + t_r \varepsilon_{,t})$, $t_r = 2D/\omega$, it can be applied to fundamental dynamic problems of continuous media. Verruijt shows with this reversible hysteretic constitutive model, using Fourier transform techniques and avoiding singularities in a smart way, that dynamic surface agitation by fast moving line loading on a visco-elastic half-plane is significantly reduced by hysteretic damping to about two times the identical static effect. This result is important for the design of high speed railway lines, constructed on soft soils.

Geostatics

In his D-Sc-thesis Verruijt [13] deals with the complex variable method for the analysis of elastic two-dimensional problems (Muskhelishvili). Not only he questioned Szelagowski's solution for the stress concentration in a sharp notch (a factor 3), but also he forecasted then the introduction of statistic soil parameters with finite element methods. Recently, Verruijt has again taken up the complex variable method for tunneling problems [14]. Though the method can cover only elasticity, it will certainly contribute, since the field effect on tunnels in soft soils is not yet fully understood.

Beside his interest in complex deformation problems attention is also given to more practical problems in foundation engineering. As a expert Verruijt is often asked to advise in juridical conflicts. Building damage, particularly for monumental structures, is mostly related to the foundation. That uniform subsidence may also cause such damage, is not to believe. Yet, in relation with corresponding groundwater-table adjustments and recalling Archimedes law, taking into account the statistical variation of natural soils, uniform landsubsidence may cause significant effects: the bearing capacity can locally decrease down to 50% [15].

Acknowledgement

The book is sponsored by a group of institutions and companies, who particularly enjoyed the professional efforts of professor Verruijt in various aspects: Fugro, GeoDelft, Hollandsche Beton Groep, KIVI-Geotechniek, Practische Studie, Public Works Rotterdam, Rijkswaterstaat, Shell Research, and Technical University Delft.

References

1. Verruijt A.: Theory of Groundwater Flow. Macmillan, London, 1970.
2. Bear J. and A. Verruijt: Modeling Groundwater Flow and Pollution. Reidel, Dordrecht, 1987.
3. Verruijt A.: Steady dispersion across an interface in a porous medium. Jl of Hydrology, Vol 14 (1971), pp. 337-347.
4. Verruijt A.: The Completeness of Biot's Solution of the Coupled Thermo-Elastic Problem. Jl Quart Appl Math, Vol XXVI-4 (1969), pp. 485-490.
5. Verruijt A.: Displacement functions in the Theory of Consolidation or in Thermo-Elasticity. Jl Appl Math & Physics, Vol 22-5 (1971), pp. 891-898.
6. Verruijt A.: Elastic Storage of Aquifers. In: Flow through Porous Media (De Wiest, ed.). Academic Press, New York, 1969, pp. 331-376.
7. Verruijt A.: The theory of consolidation. In: Transport Phenomena in Porous Media (Bear & Corapcioglu, ed.), Martinus Nijhoff, Dordrect, 1984, pp. 349-368.
8. Barends F.B.J. and A. Verruijt: A critical review of the relationship between fluid withdrawal and compaction. Int Conference on Compressibility Phenomena in Subsidence (Saxena, ed.), Engineering Foundation, New York, 1984, pp. 347-362.
9. Verruijt A.: A simple formula for the estimation of pore pressures and their dissipation. Jl Appl Ocean Research, Vol 2 (1980), pp. 57-62.
10. Vermeer P.A. and A. Verruijt: An accuracy condition for consolidation by finite elements. Jl of Num & Anal Methods in Geomechanics, Vol 5 (1981), pp. 1-14.
11. Verruijt A.: Computational geomechanics. Kluwer, Dordrecht, 1995.
12. Verruijt A.: Dynamics of soils with hysteretic damping. Proc 12th ECSMGE (Barends et al, ed.), Vol 1 (1999), pp. 3-14.
13. Verruijt A.: Stresses due to gravity in an elastic half-plane with notches or mounds. D-Sc thesis (TUDelft), Drukkerij Wed. G. van Soest, Amsterdam, 1969.
14. Verruijt A.: Deformations of an elastic half-plane with a circular cavity. Jl Solids & Structures, Vol 21 (2000), pp. 2795-2804.
15. Verruijt A.: Schade bij gelijkmatige bodemdaling. Geotechniek, Vol 4 (2000), pp. 15-18.

The Faculty of Civil Engineering and Geosciences

DELFT UNIVERSITY OF TECHNOLOGY
www.citg.tudelft.nl

Education

Activities of the faculty of Civil Engineering and Geosciences are related to

- *Geometry of the earth,*
- *Natural resources, and*
- *Built-up environment and its use.*

The purpose of our activities is the sharing specific knowledge and creation of general knowledge, that is in terms of space and time and at an international scientific level.

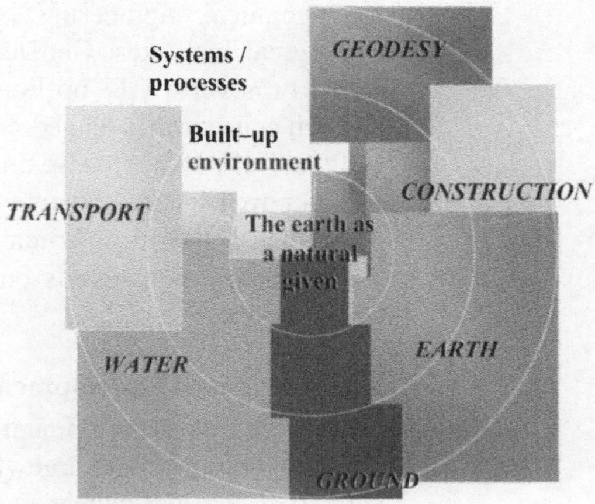

The Faculty offers educational programs for B-Sc and M-Sc in: Civil Engineering, Geodesy, and Applied Earth Sciences. The presence of graduate students gives the faculty the opportunity to do curiosity driven research surveys at relatively low costs.

Research

Research in the faculty CiTG is geared toward exploring and explaining phenomena dealing with the sustainable use of the earth and its natural resources, the functionality of the infrastructure and the (built-up) environment and the efficient use of its facilities. Research programmes are grouped into six research topics: *GEODESY, CONSTRUCTION, EARTH, GROUND, WATER* and *TRANSPORT*. This description indicates not only the connection with the individual programmes but also the technological and societal relevance of the research, often supported by national and international collaborative ventures.

A special quality of the CiTG faculty is the extensive breadth of research and educational areas. It concerns not only the subjects dealt with, but also various types of disciplines: mathematical research is conducted next to high-level experimental work, precision-mechanics and planning research. Close co-operation with technical institutions allows the faculty to maintain experimental facilities at a high scientific level. The faculty's activities respond to current and future societal topics: there is great demand for its knowledge.

GeoDelft, Partner in Risk Management

www.geodelft.nl

GeoDelft, founded in 1934, is one of the world's oldest institutes of geotechnical engineering. Nowadays, as one of the national Grand Technological Institutes, GeoDelft houses about 300 employees and has a annual turnover of 18 million Euro. It has a subsidiary company in the UK which concentrates on geo-ecological issues. One of the unique selling points of GeoDelft is the intense mix of routine work, experience based consultancy and knowledge development. Mutual fertilizing of these three levels of operational activities is crucial as a discriminating factor with respect to the geotechnical consultants on the one hand and the (technical) universities on the other.

Research and Development

Research and Development at GeoDelft are carried by the philosophy of the Innovation Cycle: knowledge development takes place, triggered out of practice, in a permanent feedback loop starting with conceptual and mathematical modeling. Subsequently, results are applied in consultancy while subject to field verification. For significant discrepancies, analysis is made of the causes, and physical modeling is applied in order to understand the mechanisms. Supported by physical model experiments, mathematical models are reformulated, which closes the innovation cycle. One cycle can run over several years and it concerns many consultancy and research projects.

Centrifuge modeling of a tunneling process.

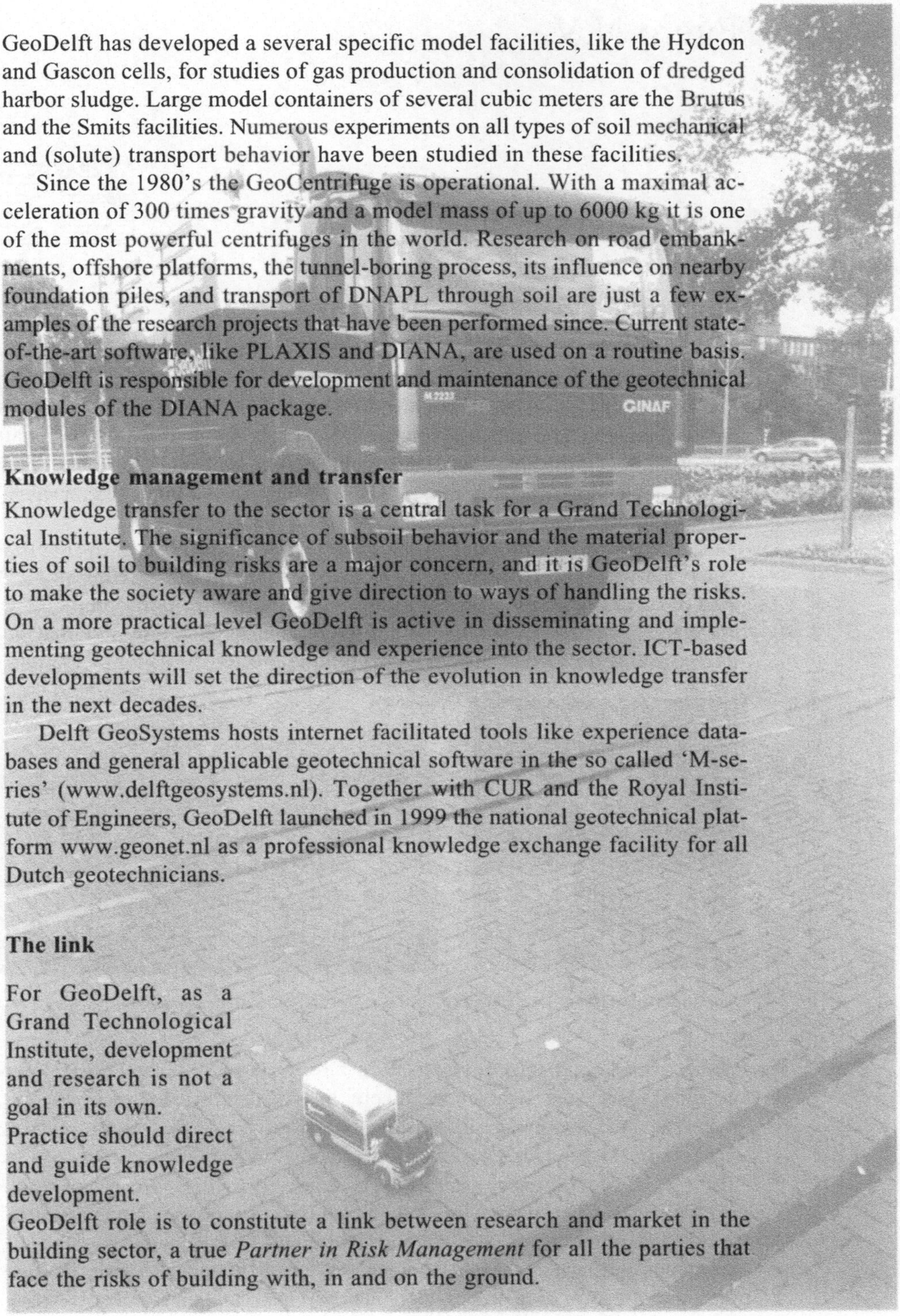

GeoDelft has developed a several specific model facilities, like the Hydcon and Gascon cells, for studies of gas production and consolidation of dredged harbor sludge. Large model containers of several cubic meters are the Brutus and the Smits facilities. Numerous experiments on all types of soil mechanical and (solute) transport behavior have been studied in these facilities.

Since the 1980's the GeoCentrifuge is operational. With a maximal acceleration of 300 times gravity and a model mass of up to 6000 kg it is one of the most powerful centrifuges in the world. Research on road embankments, offshore platforms, the tunnel-boring process, its influence on nearby foundation piles, and transport of DNAPL through soil are just a few examples of the research projects that have been performed since. Current state-of-the-art software, like PLAXIS and DIANA, are used on a routine basis. GeoDelft is responsible for development and maintenance of the geotechnical modules of the DIANA package.

Knowledge management and transfer

Knowledge transfer to the sector is a central task for a Grand Technological Institute. The significance of subsoil behavior and the material properties of soil to building risks are a major concern, and it is GeoDelft's role to make the society aware and give direction to ways of handling the risks. On a more practical level GeoDelft is active in disseminating and implementing geotechnical knowledge and experience into the sector. ICT-based developments will set the direction of the evolution in knowledge transfer in the next decades.

Delft GeoSystems hosts internet facilitated tools like experience databases and general applicable geotechnical software in the so called 'M-series' (www.delftgeosystems.nl). Together with CUR and the Royal Institute of Engineers, GeoDelft launched in 1999 the national geotechnical platform www.geonet.nl as a professional knowledge exchange facility for all Dutch geotechnicians.

The link

For GeoDelft, as a Grand Technological Institute, development and research is not a goal in its own.
Practice should direct and guide knowledge development.
GeoDelft role is to constitute a link between research and market in the building sector, a true *Partner in Risk Management* for all the parties that face the risks of building with, in and on the ground.

Fugro N.V.

www.fugro.nl

Fugro's operations focus on collecting, processing and interpreting data concerning the earth's soil and surface and the constructions built upon it, giving advice based on such data and providing related services. Fugro carries out its geotechnical and surveying services at sea, on land and from the air and provides significant added value to clients through a mix of professional, and, in the majority of cases, highly qualified staff, technologically advanced equipment and information systems.

Fugro occupies a prominent and generally leading position in international and regional markets and is steadily expanding its existing market positions. It is continuously exploiting opportunities in new, complementary niche or regional markets, very often through acquisitions, and its wide range of related services makes Fugro unique in the world.

Reconnaissance from air: Fli-map, and on sea: Scannia Carla 2001.

Organizationally Fugro operates from three divisions: Geotechnical, Survey and Geosciences. Geotechnical activities include collecting soil samples, analyzing and testing the soil and providing advice based on the results. Surveying services include ascertaining the geological composition of the earth's crust, detailing the earth's surface and providing precise positioning data. Geosciences activities include quantifying and qualifying resources for oil, gas, minerals and optimize production. All divisions use advanced equipment and systems, most of which were developed in-house, to support the collection and processing, thereby ensuring that reports are reliable and received by clients quickly. As an extension to its activities, Fugro provides services such as positioning, design advice, testing of construction materials and geographic and other information systems.

Major clients include oil and gas companies, construction companies and project developers, local government, telecommunication and energy companies, consulting engineers, architects and dredging companies. The assignments Fugro carries out on behalf of clients include surveys and services related to:

- oil and gas platforms and other constructions at sea;
- infrastructure projects (tunnels, roads, coastal works, land reclamation);
- exploration for oil and gas fields and minerals (tin, copper, gold, diamonds, etc.);
- pipelines and communication cables;
- industrial projects (buildings, factories);
- foundations and materials and environmental inspections;
- data management and geographical information systems.

Fugro, founded in 1962 and listed on Euronext N.V. in Amsterdam since 1992, has over two hundred offices throughout the world, about 7,000 staff and a permanent presence in more than fifty countries.

CPT-truck, in service around the clock.

HBG n.v.

www.hbg.nl

Learned and applied ... that's what lay behind the origins of HBG. It was exactly one hundred years ago – in January 1902 – that engineer Van Hemert founded our company. Asked by a railway company to evaluate the design for a railway viaduct in Rotterdam, Van Hemert submitted an alternative design and proposed to undertake its construction. This was the start of HBM, from which HBG later originated. Considering and proposing alternatives have remained one of the most characteristic features of the Group. A close co-operation between Delft University and HBG was guaranteed, since Van Hemert was a professor in Delft.

From the outset, our engineering skills brought us abroad, undertaking projects in places as far apart as South-America and Indonesia. At present, HBG is still operating around the world.

So many projects in so many countries over so many years ... by so many civil engineers from Delft University. HBG recognizes the impressive contribution of Prof. Arnold Verruijt in educating and stimulating so many students in soil mechanics, which is the foundation of fundamental thinking. Our main risk area in civil engineering is in soil mechanics. HBG could always appeal to Prof. Verruijt's knowledge and experience in solving complex issues.

Every engineering project is unique, every constructive solution is a unique solution as well. Therefore, it's *verruit uniek* what many of the former students of Prof. Verruijt have accomplished, also on behalf of our Group. They have learned and applied. And – even more important – they realize we all have to keep learning.

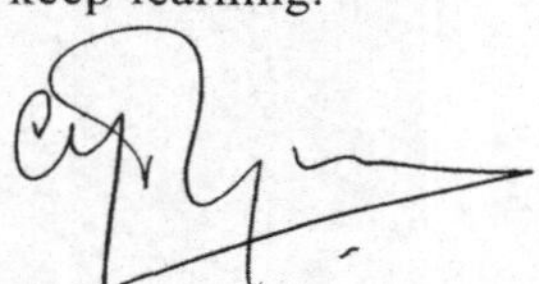

Rijswijk, 25 January 2002

Carel-Jan Reigersman
President, Board of Management
Hollandsche Beton Groep nv

Rotterdam Public Works

ENGINEERING CONSULTANTS
www.gw.rotterdam.nl

Consultancy, development and implementation of plans for infrastructure, construction and the environment are prime activities of Rotterdam Public Works Engineering Consultants. With long experience in addressing the needs of the world's largest port and a region undergoing rapid development, this highly qualified organization transforms ambitions and plans of clients into tangible results.

Engineering Consultants services vary from soil investigation to full realization of large-scale infrastructure projects, for public and private clients.

Challenge: railway underground without disturbing the monumental church.

Ministry of Transport, Public Works and Water Management

DIRECTORATE-GENERAL FOR WATER AFFAIRS
www.minvenw.nl

The Netherlands lives with the water

Life in the Netherlands means life with the water. Water is ineluctable in our landscape and inevitable for nature, a vital transportation facility and a crucial element for drinking and industry. Agriculture cannot without, and the watersides attracts many to settle and recreate. The climate is changing, sea levels rise and land subsides. This has drastic consequences for safety, water management and country planning. It is important to safeguard the best water as inheritance to our offspring, to wit, in a proper ecological constitution.

The government must define a widely supported water policy in order to guarantee and maintain an acceptable level of safety and life conditions. The expected climate change will demand, however, radical adjustments to river and coastal structures. An intensive co-operation is required in the sectors water, nature, landscape, and town and country planning to realize adequate measures and achieve the best coherence and quality.

In the Ministry of Transport, Public Works and Water Management a new Directorate-General for Water Affairs has been established in January 2002. Its main task is to protect against inundation and to accomplish sustainable use of water and its infrastructure. GD-Water is responsible for the establishment of a national water policy. It will unite, guide and control parties that execute the related plans. The options for success are interrelation of interests, collectivity in aims and intensive involvement of governmental and social partners. The result is a generally accepted and practicable frame of execution. DG Water has the ambition, with expertise and innovative zest, to give direction to the solution of water related problems of the near future.

GROUNDWATER FLOW

Moving Interfaces

Cees van den Akker, Willem-Jan Plug
TU Delft, Delft, The Netherlands

The well-known and economically important cultivation of flowers takes place in the western part of the Netherlands, alongside the dune area. Due to a number of phenomena like excessive rainfall, diminishing groundwater extraction in the dune area and the moving interfaces between fresh and saline groundwater, the groundwater tables are too high in the hinterland where the flowers grow. The rather dense surface water system with a fixed level is not adequate to maintain the optimum groundwater levels in between the ditches.
In the future, for ecological reasons, the groundwater levels in the dunes will be even higher. These higher levels will influence groundwater levels in the adjacent area, causing an increase in crop damage.

The hydrological problem is to find a methodology or technique causing high groundwater levels in the dune area, but low groundwater levels adjacent to the dunes. In principle this situation can be obtained by a series of extraction and injection wells into the fresh water system. However the behaviour of the transition zone between the fresh and saline water is rather important in order to obtain a sustainable solution. One can expect that on the long run underneath the extraction wells, the brackish zone increases unacceptably.

A second possibility in order to lower the groundwater tables is to extract saline water at a certain depth underneath the area where the flowers are cultivated.

Statement of the problem

The behaviour of the transition zone in between fresh and saline water under the influence of an extraction of saline water is the subject of interest. First, the movement of a sharp interface is investigated with and without saline water extraction. Finally, a transition zone is introduced and the behaviour of this zone is investigated.

Equation for two-dimensional flow

The equation for plane groundwater flow in an inhomogeneous anisotropic porous medium, where both the fluid and the porous medium are incompressible, is

$$\frac{\partial}{\partial x}\left(-\frac{\kappa_x}{\mu}\frac{\partial p}{\partial x}\right)+\frac{\partial}{\partial y}\left(-\frac{\kappa_y}{\mu}\left(\frac{\partial p}{\partial y}+\rho g\right)\right)=0 \tag{1}$$

where;

$x =$ horizontal coordinate [m]

$y =$ vertical coordinate [m]

$p =$ pressure [N/m^2]

$\kappa =$ intrinsic permeability [m^2]

$\mu =$ dynamic viscosity [kg/ms]

$\rho =$ density [kg/m^3]

$g =$ acceleration of gravity [m/s^2]

In terms of the stream function Ψ [m^2/s], equation (1) can be written as:

$$\frac{\partial}{\partial x}\left(\frac{\mu}{\kappa_y}\frac{\partial \Psi}{\partial x}\right)+\frac{\partial}{\partial y}\left(\frac{\mu}{\kappa_y}\frac{\partial \Psi}{\partial y}\right)=-\frac{\partial \gamma}{\partial x} \tag{2}$$

where $\gamma = \rho g$ (kg/m^{2}s^2).

$\partial \gamma / \partial x$ can be seen as a sink or source term with strength (van den Akker [1])

$$Q = (\rho_1 - \rho_2) g \Delta h \tag{3}$$

Computations at sharp interfaces

To compute interfaces a model has been developed in Matlab (Schaars et al., [2]. This model is based on the groundwater program Modflow. In this finite difference program, the groundwater heads in each cell are computed by solving the water balance of each cell in the model. An important variable is the conductance term $C_{i,j+\frac{1}{2}}$, which is written as the reciprocal of the resistance on the cell boundary between two cells (McDonald & Harbaugh[3]:

$$C_{i,j+\frac{1}{2}} = \frac{\Delta y \Delta z}{\Delta x} K_{i,j+\frac{1}{2}} \tag{4}$$

Due to the analogy between equation (1) and (2) , the values of the stream function can also be calculated by solving the water balance of each cell in terms of the stream function using the conductance term from (4) as follows

$$C_{i,j+\frac{1}{2}} = \frac{\Delta y \Delta z}{\Delta x} \frac{1}{K_{i,j+\frac{1}{2}}} \tag{5}$$

Originally, the input of the model is given as a block-centered model but, by using the stream function, it is more sophisticated and efficient to calculate the values Ψ of the stream function on the cell nodes, by introducing a mesh-centered model. For computing the discharges over the cell boundaries there will be a switch again from mesh-centered to a block-centered model. This is shown by the definition of the specific discharges q_x and q_y in terms of the stream function

$$q_x = -\frac{\partial \Psi}{\partial y} \text{ and } q_y = \frac{\partial \Psi}{\partial x} \tag{6}$$

According to equation (2) and introducing the hydraulic conductivity K [m/s] it follows that the sink or source terms can be written as

$$Q = \frac{\Delta \rho}{\rho_f} \Delta h \tag{7}$$

Using the stream function will complicate modeling extraction and injections, because they act like a vortex. Thus for each sink or source in the model area, a branchcut must be introduced (van den Akker, [1]). In this paper however, for simplicity reasons, the extractions and injections will be assumed to act on the boundary of the flow domain. When the initial and the boundary conditions are known in the model, the following equation will be solved:

$$\frac{\partial}{\partial x}\left(\frac{1}{K_y}\frac{\partial \Psi}{\partial x}\right) + \frac{\partial}{\partial y}\left(\frac{1}{K_x}\frac{\partial \Psi}{\partial y}\right) = -\frac{\Delta \rho}{\rho_f}\Delta h \tag{8}$$

By using the stream function for density driven flow the influence of dispersion and diffusion is not taken into consideration and the water flow causes the movement of the interfaces. In modeling the movement of sharp interfaces correctly, one has to use moving grids and it is advantageous to use a finite element scheme.

By the use of the finite difference method, numerical dispersion will be introduced, because it is impossible to fit the cell in such a way that the interface is placed on the cell boundaries. For that reason a procedure is introduced to attribute to a cell the density for fresh, brackish or saline water. After each time step the equations of conservation of mass (water and salt) and the calculated flow field will lead to a new position of the interfaces between fresh/brackish and brackish/salt.

The difference in specific discharges on each side of the interface and parallel to the interface can be expressed as (De Josselin de Jong, [4])

$$\Delta q_m = \frac{\kappa}{\mu}(\rho_2 - \rho_1) g \sin\alpha \tag{11}$$

In steady state situations with respect to the position of the interfaces, this relation can be used to check the numerical results (van den Akker, [1]).

Numerical results

The schematization of the initial problem with a sharp interface between fresh and saline water is indicated in Figure 1.

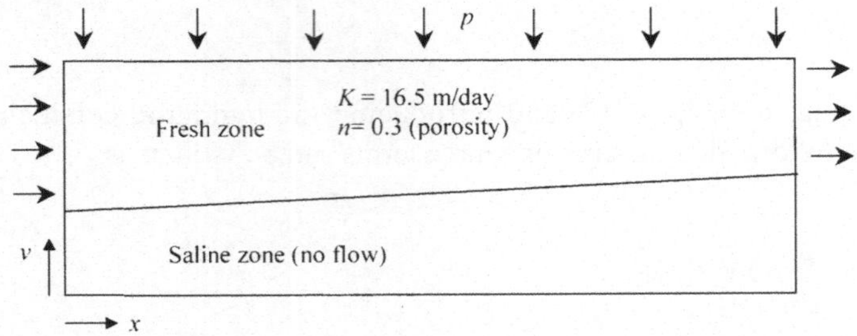

Figure 1. Schematization of the flow field with one sharp interface

- The dimension of the flow field is 400 m (horizontal) and 70 m vertical ρ_f = 1000 kg/m^3 and ρ_s =1025 kg/m^3
- The recharge on the top boundary is p = 0.001 m^2/day
- Initially there is no flow in the salt water zone
- In the fresh water zone the inflow is 1000 m × 0.001 m/day = 1 m^2/day
- Initially the slope of the interface is 1:20
- Initially at x= 0 m, the thickness of the fresh water zone is 50 m
- For the simulations, the flow model is divided in 2800 cells, with Δx =10 m and Δy =1 m
- The position of the interfaces on the left and right hand side boundaries of the flow domain is kept constant

For a number of situations, simulations were made for a period of 3600 days and an extraction of Q_{well} = 2 m^2/day on the top boundary, respectively on the lower boundary.

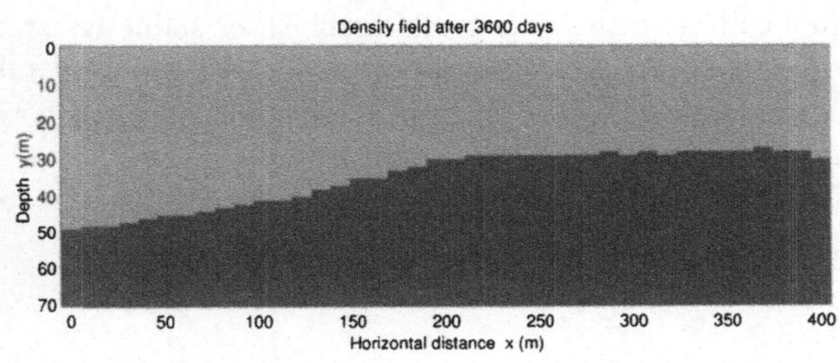

Figure 2. Density field with an extraction on the top boundary at x = 200 m

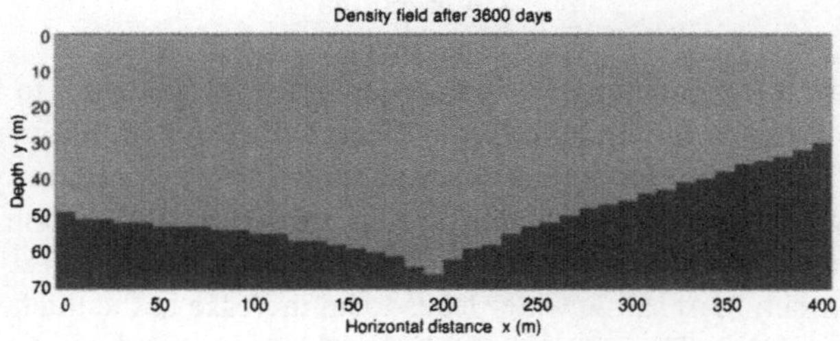

Figure 3. Density field with an extraction on the lower boundary at $x = 200$ m

The schematization of the flow field with a brackish zone, is indicated in Figure 4. Initially the thickness of the transition zone is 5 m and at x =0 m the thickness of the fresh water zone is 50 m. The density of the brackish water is 1012.5 kg/m^3

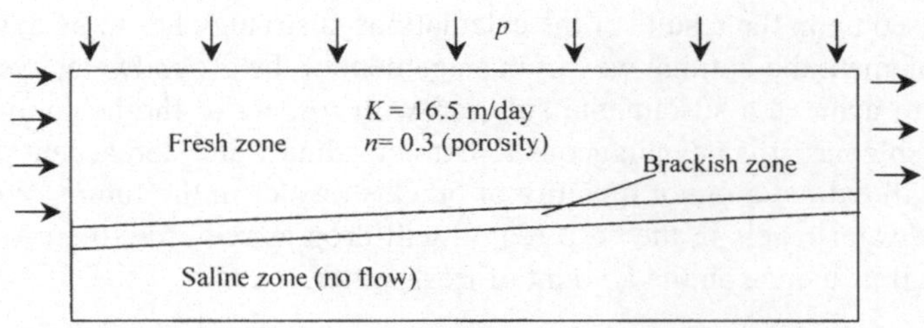

Figure 4. Schematization of the flow field with two sharp interfaces

Again the extraction Q_{well} = 2 m^2/day is placed on the top boundary, respectively on the lower boundary.

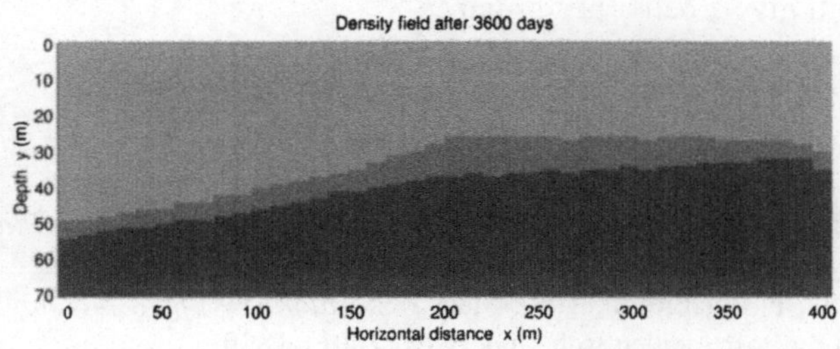

Figure 5. Density field with an extraction on the top boundary at $x = 200$ m

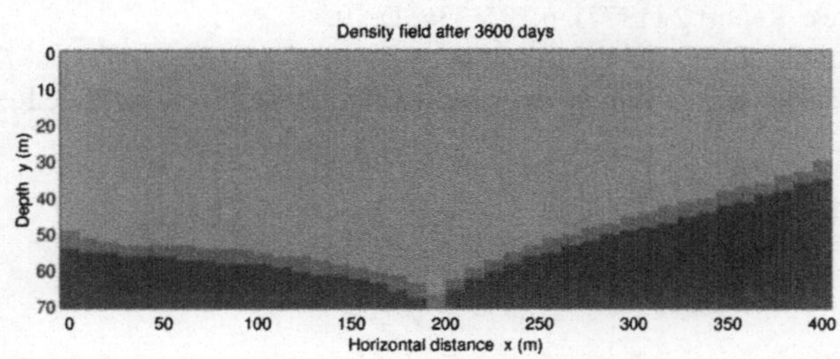

Figure 6. Density field days with an extraction on the lower boundary at $x = 200$ m

Conclusions

According to the simulations, the extraction of fresh water leads to an increase in the thickness of the brackish zone. This will result in an unsustainable situation and finally the extraction well will extract a certain quantity of brackish water. In the case where water is re-used for infiltration, or for the production of drinking water, this is not acceptable.

The extraction of saline water leads to an increase in the thickness of the fresh water pocket. The seepage of saline and brackish water to the hinterland will diminish. If the extraction rate of the well is such that a small amount of brackish water is extracted, the thickness of the brackish zone will decrease. Since the saline water from the well is not suitable for infiltration purposes or drinking water production, this quantity of brackish water is of no consequence. Transportation to the sea is probably the best solution to remove the saline water.

Based upon the results of the calculations, a strategy has to be developed to implement the optimal way of management of the groundwater system. In order to achieve a sustainable solution with respect to the behavior of the brackish zone, it is advantageous to extract saline water and accept that this well will extract a minor quantity of brackish water in the future. While the groundwater levels in the crop region will drop as required, there will be a minimal influence on the amount of fresh water.

Recommendation

Research must be continued in order to find the optimal solution in managing the fresh/brackish/saline water system. It is expected that extraction by means of wells in the fresh water system, combined with extractions in the salt water system will give a better performance.

References

1. Van den Akker, C.: *Numerical Analysis of the Stream function in Plane Groundwater* Docter Thesis. TUDelft, Delft, 1982.
2. Schaars, F., Olsthoorn, T.N., Maas, K.: *Maak Uw Eigen Modflow in Matlab*. Stichting Post Academisch Onderwijs (2001), Delft
3. De Josselin de Jong, G.: *Review of Vortex Theory for Multiple Fluid Flow*. Delft Progress Report 2 (1977), p:225-236, Delft
4. McDonald, D., Harbaugh, A.W.: *MODFLOW, A Modular Three-Dimensional Finite-Difference Groundwater Flow Model*, U.S.G.S. Open-File Report 83-875 1984.

Adaptive Grid Modeling of Unsaturated Groundwater Flow

JOHN M. VAN ESCH
GeoDelft, Delft, the Netherlands

This paper presents an adaptive multi grid - finite element method for modeling unsaturated groundwater flow. By refining only rough regions of the problem domain adaptive local grid refinement improves the computational accuracy and uses computer resources more efficiently, especially for three-dimensional simulations. The modeling technique is presented by stating the mathematical model for unsaturated groundwater flow first. This model is then solved numerically combining the finite element method and a multi grid technique. The performance of the resulting algorithm will be illustrated for a three-dimensional application and its two-dimensional counterpart. This example clearly indicates the relevance of modeling the real world in three dimensions.

Mathematical model

Bear and Verruijt [1] present the derivation of the governing equation that describes unsaturated groundwater flow. This flow equation combines a mass balance equation and Darcy's law and is known as the Richards equation

$$\frac{\partial}{\partial t}(nS\rho)=\frac{\partial}{\partial x_i}\left[\frac{\rho k K_{ij}}{\mu}\left(\frac{\partial p}{\partial x_j}-\rho g_j\right)\right] \tag{1}$$

In this equation pressure p [Pa] is the primary unknown, the independent variables are x [m] specifying the position in space along the Cartesian coordinate axis and time t [s]. The fluid density ρ [kgm^{-3}] may vary with the amount of solute dissolved in the liquid phase for instance, and its dynamic viscosity μ [kgm^{-1}s^{-1}] may vary with temperature. In this article however both are kept constant. Porosity n [-] and the saturated hydraulic conductivity tensor K [m^2] are assumed to be related to the soil only and therefore they vary in space only. The water content is described as the product of porosity and saturation S [-] which is related to fluid pressure and soil type. The relative permeability k [-] is related to saturation and soil type. Acceleration due to gravity is expressed by the gravity vector g [ms^{-2}] pointing downward in the third coordinate direction. Source terms are included as boundary conditions, leading to a more general approach.

Van Genuchten [5] presents a popular material model for saturation and relative permeability. For saturation his expression read

$$S = S^r + \left(S^s - S^r\right)\cdot\left[1 + \left(g_a|p|\right)^{g_n}\right]^{-g_m} \tag{2}$$

According to this description saturation varies between the residual saturation S^r [-] and saturation S^r [-] at saturated conditions. The relation holds for negative pressures and is C_0 continuous only for $p=0$. The expression for relative permeability is given by

$$k = (S^e)^{g_l}\left[1 - \left(1 - (S^e)^{1/g_m}\right)^{g_m}\right]^2 \tag{3}$$

Here relative permeability is related to saturation using the effective saturation S^e [-] written as $S^e = (S-S^r)/(S^s-S^r)$. The Van Genuchten material model contains three parameters: g_a [Pa^{-1}], g_l [-] and g_n [-] that are obtained from laboratory experiments and a fitting procedure. The missing parameter $g_m = 1-1/g_n$ will always be positive as $g_n > 1$. Figure 1 shows an alternative material model according to Haverkamp [3] for sand and clay, which will be used for simulation. It is noted that the sandy material shows a faster decrease in saturation and relative permeability, as pressures become more negative.

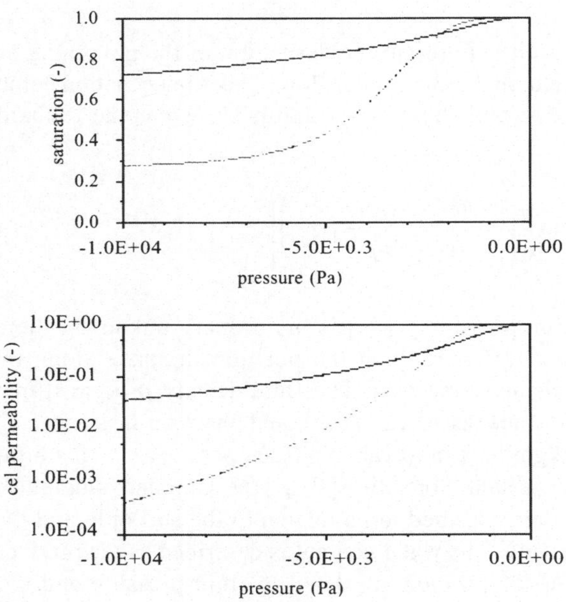

Figure 1. Pressure-saturation and pressure-relative permeability relation according to Haverkamp [3], sand (green) clay (red)

Since saturation and relative permeability are related to pressure, the initial pressure field can be found from one of both and the inverse relation. More directly the initial conditions may be given as a prescribed pressure field

$$p(x,t^0)=\bar{p}^0 \quad in \quad \Omega \tag{4}$$

Alternatively, an equilibrium condition can be calculated from a set of boundary conditions. With Ω the calculation domain, Γ its boundary, and an overline denoting a given value. Three types of boundary conditions are presented here. Dirichlet boundary conditions prescribe a given pressure as

$$p=\bar{p}(x,t) \quad on \quad \Gamma_1 \tag{5}$$

Neumann boundary conditions have no physical meaning for this problem, Cauchy conditions prescribe an out-flowing mass flux ρ

$$-\frac{\rho k K_{ij}}{\mu}\left(\frac{\partial p}{\partial x_j}+\rho g_j\right)n_i=\rho\bar{q}(x,t) \quad on \quad \Gamma_2 \tag{6}$$

Only for an in-flowing fluid the right hand side mass density is known in advance, closed boundaries are formulated with a zero right hand side. Variable boundary conditions describe seepage where outflow takes place at atmospheric pressure, or precipitation in which case ponding may occur. Both conditions can be written as

$$\begin{cases} p=\bar{p}(x,t), \quad -\dfrac{\rho k K_{ij}}{\mu}\left(\dfrac{\partial p}{\partial x_j}+\rho g_j\right)n_i>\rho\bar{q}(x,t) \\ -\dfrac{\rho k K_{ij}}{\mu}\left(\dfrac{\partial p}{\partial x_j}+\rho g_j\right)n_i=\rho\bar{q}(x,t), \quad p<\bar{p}(x,t) \end{cases} \quad on \quad \Gamma_3 \tag{7}$$

Conform this notation, the volumetric precipitation in-flux has a negative sign (q̄; < 0). The pressure p̄; at which ponding takes place will be zero for the seepage face without precipitation (q̄; = 0). Time varying boundary conditions due to a fluctuating water level may switch from Γ_1 to Γ_3 and vice versa.

As an illustration, Figure 2 shows the infiltration profiles in two homogeneous columns; the first, Haverkamp sand and the second, Haverkamp clay. Both columns initially have a constant pressure of -10^4 Pa. Infiltration takes place at atmospheric pressure applied to the top of the column. The test simulations reported by Van Esch and Spierenburg [4], and compare well. Figure 3 displays the pressure distribution for a heterogeneous situation. In the first column sand is placed on top of clay, in the second clay covers sand. Initial and boundary conditions equal the conditions for the homogeneous columns. Infiltration into the sand part of the first column takes place in the same way as in the homogeneous column until the front reaches the clay material. From that moment infiltration in the clay material occurse, continuity of pressure and flux at the interface, give rise to an increase in pressure at the interface until a near hydrostatic pressure distribution has been reached in the sand domain. For the second column, initially not only infiltration into the clay material takes place. Here a redistribution process of fluid at the interface becomes apparant. Pressure profiles are calculated using the numerical approach presented in the next chapter on a single non-adapted mesh.

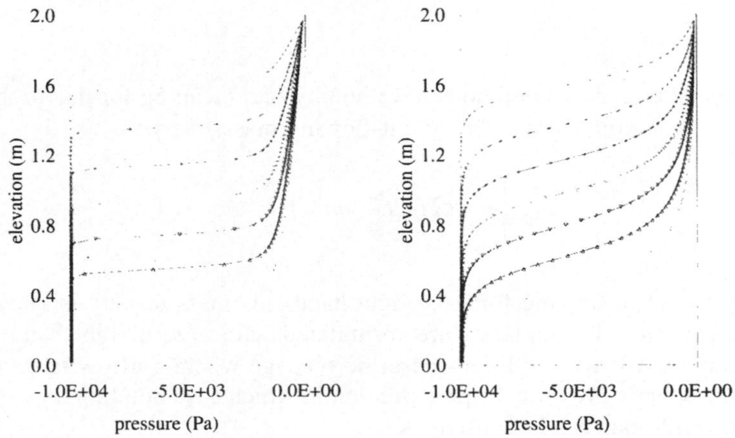

Figure 2. Pressure distribution for homogeneous profiles at infiltration, left figure sand t = 4e2, 8e2, 1.2e3, 1.6e3, 2e3, 2.4e3, 2.8e3 and 4e3 s, right figure clay t = 2e5, 4e5, 6e5, 8e5, 1e6, 1,2e6, 1.4e6 and 2.8e6 s.

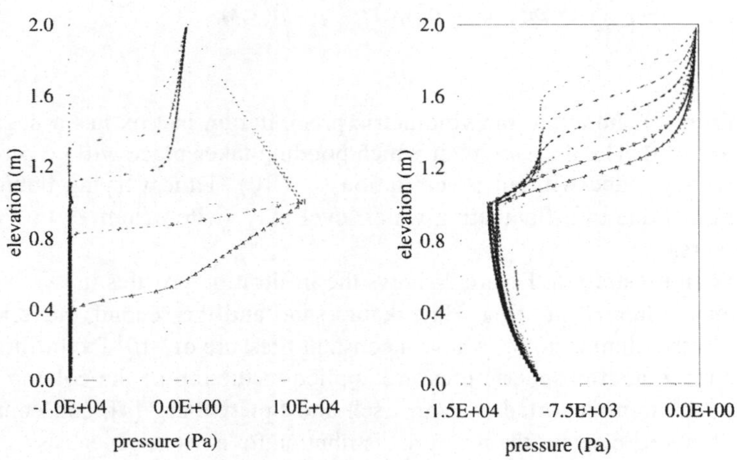

Figure 3. Pressure distribution for heterogeneous profiles at infiltration, left figure sand on top of clay t = 4e2, 8e2, 1.2e2, 1.6e2, 1e4, 1e5, 2e5 and 4e5 s, right figure clay on top of sand t = 1e5, 2e5, 3e5, 4e5, 5e6, 6e6, 7e6 and 1e6 s.

Numerical model

The finite element method is well explained by Verruijt [6] and will not be presented in detail here. In the following examples only bilinear quadrilateral elements and trilinear hexahedral elements are used to subdivide the modeling domain. However the implemented model is more general. The resulting set of non-linear equations when standard Galerkin finite element discretization and fully implicit finite difference time stepping is applied is written as

$$M_{ab}^{k+1} \frac{p_b^{k+1} - p_b^k}{\Delta t} + S_{ab}^{k+1} p_b^{k+1} = F_a^{k+1} \tag{8}$$

In the set of equations superscript k denotes the current time step. The components of the mass matrix M_{ab} are given by

$$M_{ab} = \int_\Omega N_a N_b (\alpha + n\beta) \rho S d\Omega + \int_\Omega N_a N_b n\rho \frac{S^{k+1} - S^k}{p_b^{k+1} - p_b^k} d\Omega \tag{9}$$

The second term includes the capacity matrix for water content change, its value is calculated numerically. Row sum lumping of the matrix improves the stability of the algorithm. The stiffness matrix S_{ab} reads

$$S_{ab} = \int_\Omega \frac{\rho k K_{ij}}{\mu} \frac{\partial N_a}{\partial x_i} \frac{\partial N_b}{\partial x_j} d\Omega \tag{10}$$

This contribution is found using Green's theorem for order reduction. Order reduction also contributes to the force vector F_a components

$$F_a = \int_\Omega \rho g_j \frac{\rho k K_{ij}}{\mu} \frac{\partial N_a}{\partial x_i} d\Omega - \int_{\Gamma_{2,3}} N_a (\rho \bar{q})_a d\Gamma \tag{11}$$

Integration is done numerically by Gaussian quadrate. The non-linear equations are linearized using a Picard method

$$\left(M_{ab}^{k+1,r} + \Delta t\, S_{ab}^{k+1,r}\right) p_b^{k+1,r+1} = M_{ab}^{k+1,r} p_b^k + \Delta t\, F_a^{k+1,r} \tag{12}$$

Superscript r denotes the current iterate. This process is repeated until a convergence, stated as $|p^{r+1}\text{-}p^r|\text{-}\varepsilon_1|p^1\text{-}p^0|<\varepsilon_0$, where the vector norm has been used.

Multi grid is well explained by Hackbusch [2], the method uses a grid hierarchy and is designed to solve a set of equations in $O(n)$ computational operations. For this problem, the Full Approximation Storage algorithm (FAS) is applied to resolve the non-linear set of equations generated per time step. As a criterion for refinement, the pressure gradient can be calculated. However, a better indication is found by reformulating the right hand side of Equation (1) using partial differentiation. For the resulting convection-diffusion type of equation the element Peclet number can be formulated as $Pe_i=\Delta l/k\cdot\delta k/\delta x_i$. Its value depends only on the element size Δl [m] and relative permeability. Hanging grid points are handled by linear interpolation.

Application

As an example of the adaptive technique the heterogeneous column experiment with a sand layer on top of clay was extended by imbedding the column in sand. Figure 4 shows some results obtained, the two-dimensional calculation was carried out with three grid levels, only two were used for its three-dimensional counterpart.

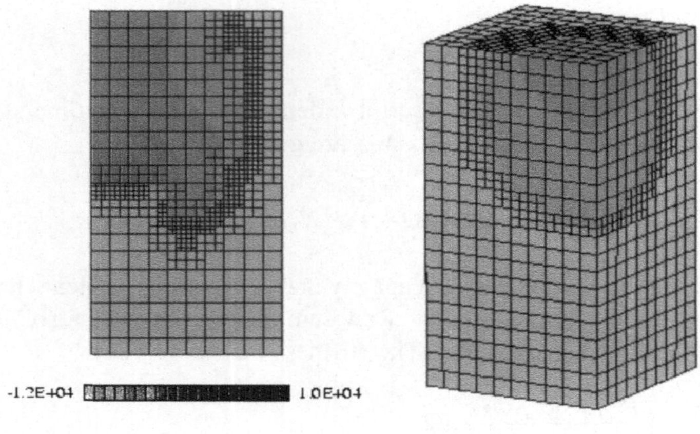

Figure 4. Pressure distribution for heterogeneous profiles t = 1600 s

The pictures show a reduction in the number of mesh nodes from 3321 to 870 in the 2D-case and 18081 to 4891 for the 3D-example. Adaptive refinement based on the element Peclet number is capable of capturing sharp interfaces thereby increasing the accuracy of the solution. The technique leaves the mesh unchanged in the rest of the modeling domain thereby ensuring computational efficiency. The difference between the results of the two-dimensional and three-dimensional calculation shows the influence of the missing dimension if one relies on 2D modeling.

References

1. Bear J. and Verruijt, A.: *Modeling groundwater flow and pollution*, Reidel, Dordrecht, 1987.
2. Hackbush, W.: *Multi-grid methods and applications*, Springer, Berlin, 1980.
3. Haverkamp, R.: A comparison of numerical simulation models for one-dimensional infiltration, Soil science soc. Am., 1977.
4. Van Esch, J.M.: and Spierenburg S.E.J. Overslag dijken onverzadigde stroming door inhomogene profielen, Delft, 1994.
5. Van Genuchten. M. Th.: A closed form equation for predicting the hydraulic function of unsaturated soils, Journal Soil Science Society of America, 44, 1980.
6. Verruijt, A.: *Computational geomechanics*, Kluwer, Dordrecht, 1995.

Resistances Against Flow in the Dutch Profile

KEES MAAS
Kiwa Water Research, Nieuwegein, and TUDelft, Delft, The Netherlands

Dutch Profile is the traditional name of a geohydrological profile made up of a single aquifer covered by an aquitard, the latter being drained by ditches (figure 1). It predominates the low lying western part of the Netherlands. Strict control of the shallow groundwater table is of primary importance to both rural and agricultural land use in this so called polder landscape. The current section presents a simple model to simulate the course of the groundwater table, dependent on recharge, seepage head and ditch water levels. The model contains two parameters that can be interpreted as resistances against groundwater flow. Expressions are given to relate them to familiar physical and geometrical quantities.

Top view of the Dutch Profile under extremely wet conditions (Aerophoto-Schiphol)

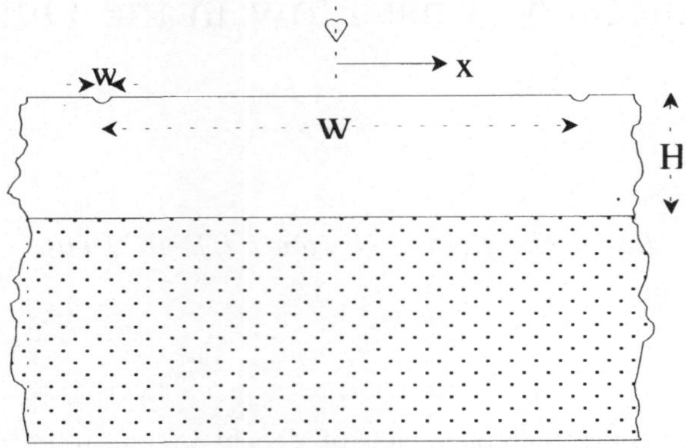

Figure 1. Dutch Profile. The ditches are assumed to run straight, parallel and equidistant

Model

Past experience with time series analyses in the given geohydrological setting indicates that the groundwater table can generally be modeled satisfactorily by a first order auto-regressive transfer function without moving average terms. This finding inspired me to model the rising and falling of the groundwater table in an arbitrary point x (Fig. 1) as a first order process, depending on the height of the water table with respect to the ditches, on recharge from precipitation, and on seepage head, according to

$$S\frac{dh}{dt} = \frac{p-h}{r_d} + R + \frac{\varphi_s}{r_s} \tag{1}$$

Here, S *[-]* = phreatic storativity, h *[m]* = groundwater table, p *[m]* = water level of the polder ditches, r_d *[d]* = constant of proportionality, which I propose to call *resistance against drainage*, R *[m/d]* = recharge from rainfall, r_s *[d]* = another constant of proportionality, which I call *resistance against seepage*, and φ_s *[m]* = seepage head in the aquifer. Seepage head is not the actual groundwater head of the aquifer; it is the part of it that causes the discharge of the ditches to exceed the amount that can be explained from recharge alone. The surplus discharge is the seepage flux s *[m/d],* integrated over the width W *[m]* (Fig. 1).

Resistances to groundwater flow

As there exists confusion about the meaning of various resistance terms that can be distinguished in the Dutch Profile, I felt free to introduce terms of my own, along with definitions. What do I mean by resistance against drainage (r_d) and resistance against seepage (r_s) ?

Not the *hydraulic resistance r [d]* of the aquitard, in any case. The hydraulic resistance is a physical property well defined in the literature as the ratio between the aquitard's thickness *H [m]* and its (vertical) permeability k_v *[m/d]*:

$$r = H/k_v \quad (2)$$

By *resistance against drainage* r_d I mean the ratio between the groundwater table h, measured with respect to the ditch water level at some point x between the ditches (Fig. 1), and the *sum* of recharge R and seepage flux s under *steady* flow conditions:

$$r_d = (h - p)/(R + s) \quad (3)$$

In this definition R and s are supposed to be uniformly distributed. As h varies with the distance to the ditches, while p, R, and s are constants, it follows that my resistance against drainage is a function of x.

I define *resistance against seepage* r_s as the ratio between the seepage head φ_s and the seepage flux s:

$$r_s = \varphi_s / s \quad (4)$$

This resistance would emerge, for instance, from a steady state pumping test.

It is evident that r_d and r_s depend on the resistance r of the aquitard, but the distance W and the width w of the ditches play a part, too, as well as the horizontal permeability k_h of the aquitard.

Later in this paper - in the appendix, that is - mathematical expressions are derived for r_d and r_s , resulting in

$$r_d = \frac{W}{\pi\sqrt{k_h k_v}} \ln(\cos\frac{\pi x}{W} / \sin\frac{\pi w}{2W}) \qquad |x| \leq (W - w)/2 \quad (5)$$

$$r_s = r - \frac{W}{\pi\sqrt{k_h k_v}} \ln(2\sin\frac{\pi w}{2W}) \quad (6)$$

Course of the groundwater table

Equation (1) has a variety of applications. An obvious example is to simulate the course of the groundwater table. To that end I reformulated (1) as

$$\frac{h}{r_d} = -S\frac{dh}{dt} + R + \frac{\varphi_s}{r_s} + \frac{p}{r_d} \quad (7)$$

The last three terms are fluxes, that can be taken together as one source term Q [m/d]:

$$\frac{h}{r_d} = -S\frac{dh}{dt} + Q \tag{8}$$

In order to compute the response of the groundwater table to an arbitrary time series *Q(t)*, it is convenient to know the *impulse response* of the groundwater table, which is the continuous analogue of the transfer function in time series models. It is also the response to a sudden recharge event of intensity 1 *[1/d]*. For such an event (8) reads

$$\frac{\theta}{r_d} = -S\frac{d\theta}{dt} + \delta(t) \tag{9}$$

I introduced the symbol θ *[-]* in stead of *h*, because the impulse response differs from *h*, by its being dimensionless. Equation (9) can be solved to give

$$\begin{aligned} \theta(x,t) &= \frac{1}{S}\exp(-\frac{t}{r_d S}) \quad (t \geq 0) \\ \theta(x,t) &= 0 (t < 0) \end{aligned} \tag{10}$$

According to a well known mathematical principle attributed to Duhamel, the response of the groundwater table to an arbitrary time series *Q(t)* can now be found by convolution:

$$h(x,t) = Q(t) \otimes \theta + h_o S\theta \tag{11}$$

The second term in the right hand side of (11) accounts for the fact that the groundwater table starts at h_0. With contemporary mathematical software, convolution is a routine operation. In Matlab, for instance, convolution of two vectors Q and θ is achieved by the command *conv(Q,θ)*.

Sample application

Figure 2 gives a sample application. It is a reconstruction of the course of the groundwater table at a location that had never been monitored. The site is subject of a lawsuit by a bankrupt pig breeder against the operator of a nearby waste disposal site. The farm lies at a short distance from a river that, being regulated by weirs, looses water to its surroundings. The transmissivity of the aquifer, the resistance against seepage and the thickness of the aquitard were

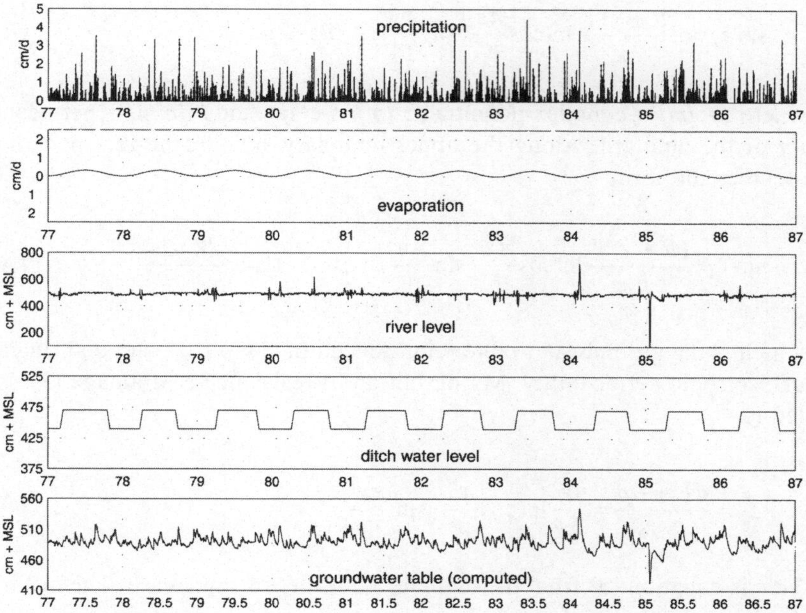

Figure 2. Sample application.

available. The distance W and the width w of the ditches were obtained from maps and from field observations. The seepage head at the site of the farm, induced by the river, was calculated by Mazure's formula:

$$\varphi_s = (h_r - p)\exp(-d / \sqrt{Tr_s}) \tag{13}$$

where h_r = river stage, d = distance to the river, T = transmissivity of the aquifer. The anisotropy of the aquitard (k_h/k_v) had to be guessed, in order to calculate the resistance against drainage from (5) and (6).

Appendix: Derivation of equations (5), (6) and (12)

The main resistance against flow builds up at the top of the aquitard, near the ditches. It is important to model that part of the flow region more or less accurately. The conditions along the lower boundary of the aquitard are not so formidable and allow for a cruder approximation. To start with, I supposed the aquitard to be infinitely thick, with the seepage flux originating from great depth. These approximations turned the flow domain between the ditches into a semi infinite strip, which could be mapped conformally onto a lower half plane In the auxiliary plane the form of the complex potential was recognized by inspection. Inverse transformation from the auxiliary plane to the physical x,y-plane gave

$$\Omega(z) = \frac{(R+s)W}{\pi k} \ln(\cos\frac{\pi z}{W} / \sin\frac{\pi w}{2W}) + iRz + p \qquad \text{(A1)}$$

where Ω *[m]* = complex potential, z *[m]* = $x+iy$, and k *[m/d]* = permeability of the aquitard. Along the upper boundary $(y = 0)$ the real part of Ω assumes the form

$$h(x) = \frac{(R+s)W}{\pi k} \ln(\cos\frac{\pi x}{W} / \sin\frac{\pi w}{2W}) + p \qquad (|x| < \frac{W-w}{2}) \qquad \text{(A2)}$$

which is the groundwater table h introduced in (1) in the main text, in case of isotropic permeability. My definition of resistance to drainage (3) now led to

$$c_d = \frac{\varphi(x) - p}{R+s} = \frac{W}{\pi k} \ln(\cos\frac{\pi x}{W} / \sin\frac{\pi w}{2W} \qquad \text{(A3)}$$

This is equation (5) from the main text, for the isotropic case. The mean of $\Omega(z)$ at a depth $y = -H$ could be shown to be

$$\overline{\Omega} = \frac{(R+s)W}{\pi k} \ln(2\sin\frac{\pi w}{2W}) + \frac{sH}{k} + p \qquad \text{(A9)}$$

which is a real number. I assumed this value to correspond well to the of head of the aquifer. In the main text, the seepage head φ_s was defined to be the part of the head of the aquifer that is responsible for the seepage flux s

$$\varphi_s = -\frac{sW}{\pi k} \ln(2\sin\frac{\pi w}{2W}) + \frac{sH}{k} \qquad \text{(A10)}$$

The resistance against seepage now followed from the definition

$$r_d = \frac{\varphi_k}{s} = \frac{H}{k} - \frac{s}{\pi k} \ln(2\sin\frac{\pi w}{2W}) \qquad \text{(A11)}$$

This is equation (6) from the main text, for the isotropic case. The isotropic results were extended to anisotropic aquitards by replacing k by $(vk_h\ k_v)$,while multiplying the horizontal dimensions by (vk_h/k_v), where k_h = horizontal permeabiltiy and k_v = vertical permeability.

Approximate Modeling of Curved Bounding Equipotentials by the Hodograph Method

OTTO D.L. STRACK
University of Minnesota, Minneapolis, Minnesota

Problems with free boundaries, such as a phreatic surface, may be solved using the hodograph method, see Hamel [1]. Applications of the hodograph method are restricted to certain types of boundary conditions (e.g.,[2], [3]), which do not include fixed curved boundaries. An approximate method is presented in this paper for including curved bounding equipotentials. The method is useful for cases where the precise shape of the boundary is not a concern.

The method will be applied to a modification of a problem solved by Verruijt [4], which concerns unconfined flow with rainfall between two horizontal bounding equipotentials. The problem solved in this paper differs from the latter in that one of the boundaries is the curved bottom of a lake, approximated by a curve with a shape controlled by a single parameter.

The background of the problem is the design of a waste repository near the bottom of a hillside with a lake higher up the hill. A drain will be placed below the repository in order to keep it dry. The objective of the analysis is to obtain an expression for the width of the drain required to achieve a given dry length of the drain.

The problem is illustrated in figure 1; the drain is section 4-5, and the dry section is labeled as 3-5. The phreatic surface below the hillside is section $\mathcal{S}$-3, and the lake bottom is approximated as the curve $\mathcal{RS}$. Vertical planes of symmetry are assumed through the center of the drain, 1$\mathcal{B}$-5, and the center of the lake, 1$\mathcal{A}$-$\mathcal{R}$. The flow is steady and occurs only in the vertical plane. The aquifer is infinitely deep with uniform hydraulic conductivity k [L/T].

The technical description of the boundary of the flow domain is as follows, see figure 1. 1$\mathcal{A}$-$\mathcal{R}$ is a streamline, $\mathcal{RS}$ is an equipotential, $\mathcal{S}$-3 is a phreatic surface with infiltration at a rate N, 3-4-5 is an equipotential and 5-1$\mathcal{B}$ is a streamline. The area of interest is the drain and its neighborhood; the presence of the lake greatly influences the groundwater flow in this case, but the precise shape of the lake bottom less important. This justifies the approximation shown in figure 2: the streamline 1$\mathcal{A}$-$\mathcal{R}$ and the phreatic surface $\mathcal{S}$-3 are both extended upward to infinity, and the water supplied by the lake is generated by a source of recharge Q_0 [L^2/T] at infinity. The discharge Q_0 is adjusted such that the phreatic surface passes through a fixed point $\mathcal{S}$, and the bottom of the lake is replaced by the equipotential $\mathcal{RS}$.

The advantages of this approximation are two-fold. First, there does, at present, not exist any technique for dealing with curved boundaries such as the river bottom $\mathcal{RS}$ in the hodograph method; the lake bottom would need to be

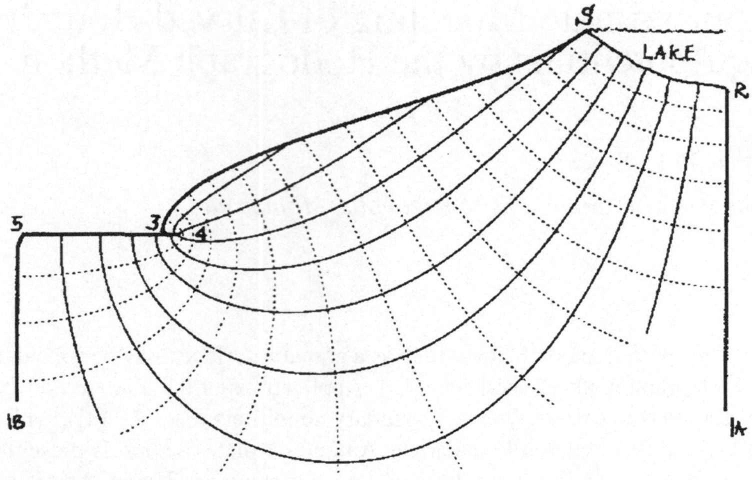

Figure 1. The actual flow domain

approximated by a horizontal line. Second, even with the latter approximation, the solution would be far more elaborate than the one presented here.

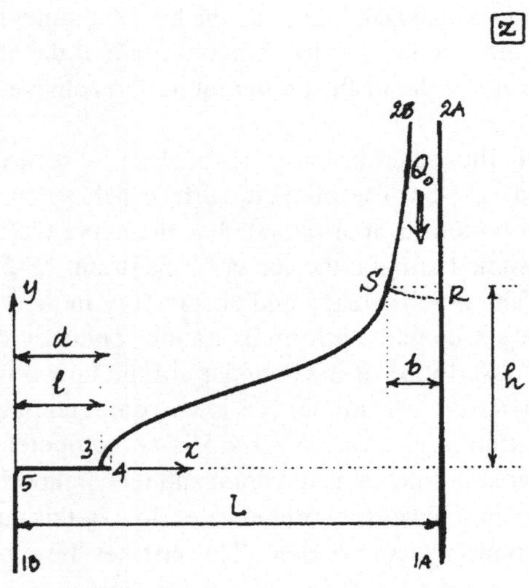

Figure 2. The approximate flow domain

As the title of this paper suggests, the proposed method for approximating the effect of lakes or streams intersecting a phreatic surface is not limited to the present application, but may be used whenever the stream or lake is recharging an aquifer along its entire bottom.

Method of solution

The problem is solved by the use of the two auxiliary functions $\overset{1}{T}$ and $\overset{2}{T}$ introduced in [5] and [3]. The reader is referred to the latter reference for detailed description of the method.

The complex variable $z = x + iy$ is defined in the physical plane, with the origin $z = 0$ at point 5 and the x axis along the drain. Use is made of the complex potential $\Omega = \Phi + i\Psi$, where $\Phi = kB\phi$ is the potential; ϕ represents the piezometric head [L], k the hydraulic conductivity [L/T], and B [L] the width of the flow domain normal to the plane of drawing. The imaginary part of Ω, Ψ, [L^3/T] is the stream function. The complex potential Ω is an analytic function of z, and will be determined by the use of conformal mapping techniques.

The distance between the two vertical impermeable boundaries, $1\mathcal{A}$-$2\mathcal{A}$ and 5-$1\mathcal{B}$ is L, the length of the drain is d, and the dry length of the drain is l. The width of the lake is $2b$, and the elevation of point S above the x axis is h. The infiltration from rainfall along $2\mathcal{B}$-3 is N.

The conditions along the boundaries of the flow domain may be expressed in terms of Φ and Ψ as follows. $1\mathcal{A}$-$2\mathcal{A}$ is a streamline, Ψ= constant. Choosing $\Psi = 0$, one obtains

$$x = L \qquad -\infty < y \leq \infty \qquad \Psi = 0 \tag{1}$$

$2\mathcal{B}$-3 is a phreatic surface with infiltration rate N. The source of strength Q_0 causes a gap to exist at infinity between the vertical streamline and the phreatic surface. Taking atmospheric pressure as zero, ϕ equals y along the entire phreatic surface so that $\Phi = kBy$ there. Since the phreatic surface continues past point S, outside the actual flow domain toward infinity, Φ approaches kBy near infinity. Thus, the flow rate Q_y for $y \to \infty$ approaches $-\partial\Phi/\partial y = -kB$, so that the width of the source near infinity is $Q_0/(kB)$; the value of Ψ at point $2\mathcal{B}$ is Q_0. Hence:

$$\lim_{\substack{x \to L - b_\infty \\ y \to \infty}} \Psi = \Psi_{2\mathcal{B}} = Q_0 \qquad b_\infty = \frac{Q_0}{kB} \tag{2}$$

where b_∞ represents the width of the source near infinity. Along the phreatic surface $2\mathcal{B}$-3, with infiltration N, the boundary conditions are that the pressure is zero, so that $\Phi = -kBy$, and $\Psi = -NBx + C$ (see [3]), where C is a constant of integration, solved for by the use of (2). The boundary condition along the phreatic surface becomes

$$l < x \leq L - \frac{Q_0}{kB} \qquad 0 < y < \infty :$$

$$\Phi = kBy \qquad \Psi = -NB\left[x - \left(L - \frac{Q_0}{kB}\right)\right] + Q_0 \tag{3}$$

Along the drain 3-4-5, the head and potential are both equal to zero:

$$0 \leq x \leq d \qquad y = 0 \qquad \Phi = 0 \tag{4}$$

The remaining boundary condition is that 5-1$\mathcal{B}$ is a streamline. It is assumed that no water is lost to negative infinity, so that the values of Ψ along 5-1$\mathcal{B}$ and 1$\mathcal{A}$-2$\mathcal{A}$ must be the same:

$$x = 0 \qquad -\infty < y \leq 0 \qquad \Psi = 0 \tag{5}$$

The two auxiliary functions

The two auxiliary functions best suited to solve the present problem may be obtained by drawing the domain in the hodograph plane, and then establishing whether all boundary segments intersect (possibly after extension) at either none, one, or two points.

Both the vertical streamlines and the horizontal drain correspond to segments of the line $Q_x = 0$ in the hodograph plane, $\overline{W}$. The phreatic surface maps onto a circle of diameter $k - N$ through $\overline{W} = -NBi$ and $\overline{W} = -kBi$. Hence, all boundary segments lie either on a circle or on a line; the circle and line intersect at two points: $\overline{W} = -NBi$ and $\overline{W} = -kBi$. Strack ([3], page 531) demonstrates how these two points of intersection lead unambiguously to the following unique pair of auxiliary functions:

$$\overset{1}{T} = z = \Omega + ikBz \qquad \overset{2}{T} = \Omega + iNBz \tag{6}$$

The first of these is the Zhukovski function.

The advantage of using the functions $\overset{1}{T}$ and $\overset{2}{T}$ over using the hodograph is that the former two functions are linear combinations of Ω and z, whereas the latter one is the complex conjugate of the derivative of Ω with respect to z. Using this derivative in the conformal mapping procedures requires a step of integration to obtain Ω as a function of z, whereas such integration is avoided by using the two auxiliary functions. It may be noted that the two auxiliary functions applied here could also be used for the problem solved by Verruijt [4].

The boundaries of the domains in the planes of the auxiliary functions consist of straight line segments. The functions $\overset{1}{T}$ and $\overset{2}{T}$ will both be expressed in terms of a reference parameter $\zeta = \xi + i\eta$ defined in the upper half plane $\Im\zeta \geq 0$ shown in figure 3(a). The functions $\Omega = \Omega(\zeta)$ and $z = z(\zeta)$ can then be obtained directly from (6) as follows:

$$z = \frac{1}{iB(k-N)}\left[\overset{1}{T}(\zeta) - \overset{2}{T}(\zeta)\right] \tag{7}$$

and

$$\Omega = \frac{1}{(k-N)}\left[N\overset{1}{T}(\zeta) - k\overset{2}{T}(\zeta)\right] \tag{8}$$

The first auxiliary function

The domain in the ζ plane is shown in figure 3(a). Point 1 corresponds to infinity in the ζ plane, points 2 and 5 are at $\zeta = -1$ and $\zeta = +1$, respectively, point 3 corresponds to $\zeta = \zeta_3 = \lambda$, and point 4 to $\zeta = \zeta_4 = \xi_4$:

$$\zeta_1 = \infty \qquad \zeta_2 = -1 \qquad \zeta_3 = \lambda \qquad \zeta_4 = \xi_4 \qquad \zeta_5 = 1 \tag{9}$$

The boundary of the domain in the $\overset{1}{T}$ plane is the semi-infinite strip of width

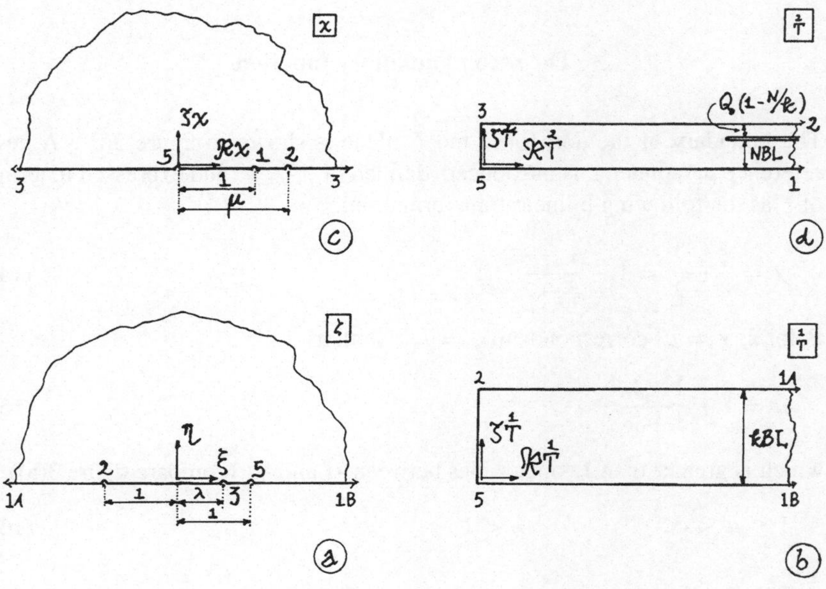

Figure 3. Boundaries in the ζ, $\overset{1}{T}$, χ, and $\overset{2}{T}$ planes

kBL shown in figure 3(b). The function that maps $\Im\zeta \geq 0$ onto the domain in the $\overset{1}{T}$ plane is

$$\overset{1}{T} = \frac{kBL}{\pi} \operatorname{arccosh} \zeta = \frac{kBL}{\pi} \ln\left[\zeta + \sqrt{\zeta^2 - 1}\right] \tag{10}$$

with the following inverse:

$$\zeta = \cosh \frac{\pi \overset{1}{T}}{kBL} \tag{11}$$

An expression for λ is obtained as follows. $\overset{1}{T}$ equals $i(\Psi_3 + kBx_3)$ at point 3. The value of Ψ_3 equals the total amount of water that enters the flow domain between points $2\mathcal{A}$ and 3 (see figure 2). This amount equals $Q_0 = kBb_\infty$ plus the water that enters the flow domain between points $2\mathcal{B}$ ($x = L - b_\infty$) and 3 ($x = l$) which equals $NB(L - b_\infty - l)$. Hence, with (2), Ψ_3 becomes

$$\Psi_3 = kBb_\infty + NB(L - b_\infty - l) = NB(L - l) + \big(1 - \frac{N}{k}\big)Q_0 \tag{12}$$

Setting $\zeta = \lambda$ and $\overset{1}{T} = i(\Psi_3 + kBl)$ in (11) gives

$$\lambda = \cos\left[\pi\left\{\frac{N}{k} + \left(1 - \frac{N}{k}\right)\left(\frac{l}{L} + \frac{Q_0}{kBL}\right)\right\}\right] \tag{13}$$

The second auxiliary function

The boundary of the domain in the $\overset{2}{T}$ plane is shown in figure 3(d). A new reference parameter χ is introduced, defined in $\Im\chi \geq 0$ and expressed in terms of ζ as the following bilinear transformation:

$$\chi = \frac{\zeta - 1}{\zeta - \lambda} = 1 - \frac{1 - \lambda}{\zeta - \lambda} \tag{14}$$

Point 2, $\chi = \mu$, corresponds to $\zeta = -1$, so that

$$\mu = 1 + \frac{1 - \lambda}{1 + \lambda} \tag{15}$$

which is greater than 1, since λ lies between -1 and +1 (compare figure 3(a))

$$1 \leq \mu < \infty \qquad -1 < \lambda < 1 \tag{16}$$

The function that maps $\Im\chi \geq 0$ onto the domain in the $\overset{2}{T}$ plane is obtained using the Schwarz-Christoffel transformation. This application of the transformation is elementary, and leads to the following result:

$$\begin{aligned}\overset{2}{T} = &-\frac{NBL}{\pi}\ln\frac{\sqrt{\chi} - 1}{\sqrt{\chi} + 1} - \frac{Q_0(1 - N/k)}{\pi}\ln\frac{\sqrt{\chi} - \sqrt{\mu}}{\sqrt{\chi} + \sqrt{\mu}} \\ &+ i[NBL + Q_0(1 - N/k)]\end{aligned} \tag{17}$$

It may be verified that (17) meets the boundary conditions shown in figure 3(d). Along 3-4-5, $\sqrt{\chi}$ is purely imaginary, and so are all terms in (17), so that $\Re\overset{2}{T} = 0$. Along 5-1, the imaginary parts of both logarithms in (17) are constant and equal to π, so that $\Im\overset{2}{T} = 0$ there. Along $1\mathcal{A}$-2, $\sqrt{\chi}$ is real and between 1 and μ. Hence, the first logarithm in (17) is real, whereas the second one has an imaginary part of π so that $\Im\overset{2}{T} = NBL$. Along 2-3, both logarithms are real, so that $\Im\overset{2}{T} = NBL + Q_0(1 - N/k)$. The boundary values of $\overset{2}{T}$ are thus in agreement with figures 3(c) and 3(d).

The functions $\overset{1}{T}$ and $\overset{2}{T}$ can now be expressed in terms of ζ; the functional relationship between the complex potential Ω and position z is obtained from (7) and (8). The relationship between the parameters of the problem on the one hand and the given boundary values and aquifer parameters on the other hand will be considered next.

Determination of the parameters

The main question to be addressed with the present solution is to find the drain length d required to achieve a dry drain length l, given N, k, and the coordinates of the lake or river edge S (see figure 1). For the sake of convenience, the functions z and Ω are replaced by the following dimensionless forms:

$$z^* = z/L \qquad \Omega^* = \Omega/(kBL) \tag{18}$$

The dimensionless parameters are λ,μ, N/k, l/L, and $Q_0/(kBL)$. Both λ and μ are expressed in terms of the latter three parameters; these are the three remaining degrees of freedom. Since N/k and l/L are given, $Q_0/(kBL)$ is the one parameter that remains to be determined. This is done by fixing point S in the dimensionless z^* plane:

$$1 - \frac{b}{L} + i\frac{h}{L} = z^*(\rho) \tag{19}$$

where ρ is a real number representing the location of point S on the real axis of the ζ plane. Point S lies on the free surface, and it follows from figure 3(a) that $-1 < \rho < \lambda$. Equation (19) is a complex equation with the unknowns ρ and Q_0/kBL, and can be solved using a Newton-Raphson iteration. Separation of $z^*(\rho)$ into real and imaginary parts yields

$$1 - \frac{b}{L} = \frac{1/\pi}{1 - N/k} \arctan \frac{\sqrt{1-\rho^2}}{\rho} - \frac{N/k}{1 - N/k} - \frac{Q_0}{kBL} \tag{20}$$

and

$$\frac{h}{L} = \frac{N/k}{1 - N/k}\frac{1}{\pi} \ln \frac{\sqrt{R}+1}{\sqrt{R}-1} + \frac{Q_0}{kBL}\frac{1}{\pi} \ln \frac{\sqrt{R}+\sqrt{\mu}}{\sqrt{R}-\sqrt{\mu}} \tag{21}$$

where R is the value of χ that corresponds to $\zeta = \rho$, or, with (14),

$$R = 1 + \frac{1-\lambda}{\lambda - \rho} = \frac{1-\rho}{\lambda-\rho} \tag{22}$$

The remaining task is to compute the length d of the drain. The drain tip, point 4, corresponds to a regular point of the real axis in both the ζ and χ planes so that

$$\frac{dz^*(\zeta_4)}{d\chi} = 0 \tag{23}$$

where $\zeta_4 = \xi_4$, and $\chi_4 = -p$ is the value of χ at point 4, which is seen from figure 3(a) to be real and negative, i.e., $p > 0$.

The derivative of z^* with respect to χ is obtained by differentiation of (7) with respect to ζ, and multiplying the result by the derivative of ζ with respect to χ, obtained from (14). This yields with condition (23) a single equation for the unknown p (the imaginary part of (23) is always zero because point 4 lies on the drain). After considerable, but elementary, elaboration, the following expression is obtained:

$$\sqrt{p+\mu} = \frac{\frac{1}{2}\sqrt{\mu}-1}{\frac{N/k}{1-N/k}+\frac{Q_0}{kBL}\sqrt{\mu}} \left[\frac{1}{1-N/k}+\sqrt{\left(\frac{N/k}{1-N/k}+\frac{2Q_0}{kBL}\sqrt{\mu}\right)^2+\frac{1+N/k}{1-N/k}}\right] \tag{24}$$

The value for ζ at point 4 is obtained from $\chi_4 = -p$ by the use of (14):

$$\zeta_4 = \xi_4 = \frac{1-\lambda\chi_4}{1-\chi_4} = \frac{1+\lambda p}{1+p} = a \tag{25}$$

Substitution of these values for λ and ζ in the expression for z^* (obtained from (7)) yields an expression for the dry drain length , after elaboration,

$$\frac{d}{L} = \frac{1}{1-N/k}\frac{1}{\pi}\arctan\frac{\sqrt{1-a^2}}{a} - \frac{2N/k}{1-N/k}\frac{1}{\pi}\arctan\sqrt{p} - \frac{2Q_0}{\pi kBL}\arctan\sqrt{\frac{p}{\mu}} \tag{26}$$

References

1. Hamel, G.: Über Grundwasser. *Z.Angew.Math.Mech.* 14 (3) (1934), pp. 129-157
2. Verruijt, A.: *Groundwater Flow.* Macmillan Press LTD, 1970.
3. Strack, O.D.L.: *Groundwater Mechanics.* Prentice Hall, Englewood Cliffs, NJ, 1989.
4. Verruijt, A.: Groundwater Flow Under a Water Divide. In: *Proceedings of the Third International Conference on the Analytic Element Method in Groundwater Flow,* Vol 2, Brainerd, MN, 2000.
5. Strack, O.D.L., and Asgian, M.: A new function for the hodograph method. *Water Resources Research* 12 (6) (1978), pp.1045-1058.

The Effects of Groundwater Flow through Grouted Layers

Frits van Tol

Rotterdam Public Works & TUDelft, Delft, The Netherlands

In regions with a high groundwater table, underground structures such as tunnels, shafts and underground parking facilities are built in dry construction pits with sheet pile or diaphragm walls to retain the groundwater and the soil. Underwater pored concrete or grouted layers are used to ensure the vertical stability of the bottom of the pit or to reduce the amount of percolating groundwater. Until about 1985, such grouted layers consisted of permeation grouting. The injected fluid was a gel, usually a soft silica-gel. From around 1985, jetgrouting was also used. With this technique, the soil is cut by jetting and mixed and partly replaced by grout. The strength and stiffness of these grout layers is comparable with a poor concrete.

The state of the art for injection technology is such that grouted layers are not impervious. This statement is the commonly-held opinion regarding permeation grouting, and is based on numerous case histories and analysis of the permeation process. Usual specifications in tender documents take this fact into account: in general, the existing permeability of the sand layer to be injected must be reduced by a factor between 10^2 and 10^3, resulting in a permeability in the range of of 10^{-7} to 2.10^{-8} m/s.

The fact, that groundwater flows through injected layers, is usually not a problem. The percolating groundwater is controlled by a drainage system in the layer above the gel layer. In the first part of this paper, the groundwater flow above the injected layer is described analytically. Next, a peculiar effect of seepage through grouted layers permeated with a soft silica-gel is presented.

The state of the art of jetgrout technology regarding the permeability that this technique can achieve is similar to that for permeation grouting. Nevertheless, various designs indicate that the perceptions of some designers differ. On several occasions, constructions with a jetgrout layer just below the excavation led to extremely serious problems when a mixture of water and soil flowed into the construction pit. Leakage through a jetgrouted arch or layer is extremely difficult to control when a thin sand layer covers the grout layer. These solutions presume a perfectly impervious jetgrout layer.

Tol [1] [2] showed that vertical jetgrouted water-retaining structures, such as a column screen or a combined wall of columns and double panels, produce a reduction in permeability ranging from 10, for the panel grouting, to 50, for the wall, and to 200, for a combined wall. The average permeability of grouted walls in a test pit was found 2.4 10^{-6} m/s.

Pumping tests carried out in the Tram Tunnel project in The Hague, where a grout arch was used, showed that the average hydraulic resistance for the different sections ranged from approximately 150 up to more then 300 days. This corresponds with a permeability between 1 10^{-7} and 2 10^{-7} m/s. A good result, but nevertheless severe problems arose when groundwater and soil flowed into the construction pit. Analysis (Tol [3]) confirms that the probability of a hole in a jetgrouted layer at a depth of 15 m below the surface, based on a probabilistic calculation using a Monte Carlo simulation, was equal to 100%.

The second part of this paper reviews the consequences of groundwater flow through a jetgrout layer. The initiation of a sand boil in the grout layer is considered first, followed by an analysis of scour under the jetgrout arch, when a sand boil is formed and a water and sand mixture flows through a hole in the arch into the construction pit.

Permeation grouting

Groundwater flow in the aquifer above a layer of permeation grouting

Layers formed by permeation grouting with soft silica-gel must be placed at a sufficient depth underneath the excavation, to guarantee the vertical equilibrium of the construction pit. The percolating groundwater can be pumped out in the sand layer above the injection layer. Elprama and Tol [4] present an analytical solution for the flow in the sand above the gel layer, both for two-dimensional and for axi-symmetrical flow. These solutions are based on the following boundary conditions: a potential line ($\varphi=\varphi_l$) at one side (a well with a fixed dewatering level), an impervious boundary along the upper side, formed by an overlaying clay layer and in the middle of the construction pit, being the symmetry axis. At the lower side the boundary condition is a layer with a uniform hydraulic resistance C, under which a constant head ($\varphi=\varphi_0$), see Figure 1.

Figure 1, taken from Elprama and Tol [4], gives according to the chosen solution equipotential lines for isotropic sand with a permeability $I = k_V = 5$ m/day and a resistance $C = 5$ days, and for anistropic sand with $k_H = 9k_V = 5$ m/day and identical resistance. As expected, it appears that the stability of the covering layer is favourably influenced by the anisotropy in the sand layer. Figure 1 shows that with the given ratio between permeability and resistance, α increases from 0.78 to 0.90, namely that the maximum head under the covering layer, in the middle of the construction pit, decreases from 22% of the pressure difference (φ_l - φ_0) to 10% of the pressure difference. Using the equations presented in the mentioned paper, the flow in the aquifer above a grouted layer with a uniform permeability can be calculated and a system of bleeder wells or any other kind of drainage system can be designed. To my knowledge, no analytical solution exists for a layer with a non-uniform resistance, e.g. holes in the grout layer.

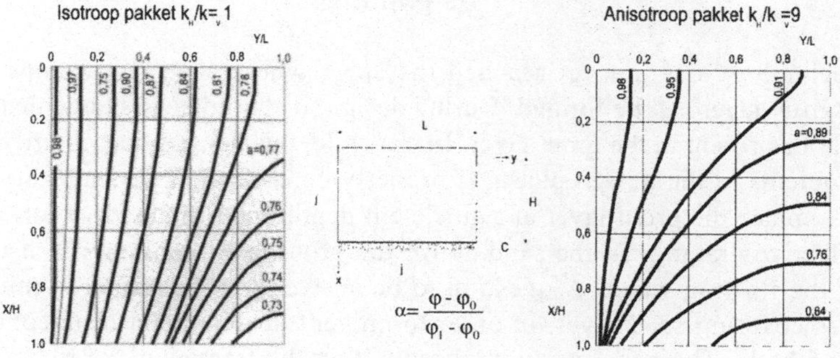

Figure 1. Equipotential lines for isotropic and anisotropic situations with boundary conditions for the 2D solution.

Groundwater flow through a grout layer of silica-gel

Groundwater flow through a silica-gel layer initiates a process, where surplus sodium hydroxide is dissolved out of the gel layer. This causes the gel layer to age, which means a slowly increasing permeability. In addition, the acidity (pH) of the groundwater flowing through the gel layer increases.

Water retaining layers as a result of permeation grouting with soft silica-gel have been performed for decades. These type of layers are only used in the construction phases of underground works, namely, to keep a construction pit dry during construction for a period of 2 up to 3 years. In general, results are positive. The rinsing process is usually so slow that the lowered resistance of the gel layer causes no problems during this construction period. Few exceptions are known. In the Tram Tunnel in The Hague, part of the construction pit was provided with a silica-gel layer. After functioning for approximately 3 years, the water-retaining function appeared to have decreased down to a constant flow rate of 0.1 m/day. The sand above the gel layer posed a problem, however, as the long-term rinsing led to substantially higher pH levels. The acidity just above the injection layer had increased to such a level that organic matter probably dissolved, and then was transported in dissolved form by the groundwater to where the pH has settled to more neutral levels.

Laboratory simulations confirm (Hoek [5]) that this can cause the permeability of the sand to decrease by a factor 50. In practice, it seems that the sand above the gel layer could no longer be drained due to this phenomena, so that pore pressures could not be controlled. After a few days, the discharge of the drainage filters reduced to a negligible flow. In case of the Tram Tunnel project, the pore pressure could only be controlled using an massive drainage capacity.

Jetgrouting

Technology has not yet reached the stage where a perfectly impervious jetgrout layer can be formed. During design, it should be assumed that holes will be present in the grout layer. In principle, this does not necessarily cause problems in the design phase, if properly considered. The simplest option is to place the grout layer at a sufficient depth under the excavation so that effective drainage in the sand above the grout layer can control discharge of the flowing water. A question to be answered is the required thickness of a covering sand layer, in order to prevent development of uncontrolled sand boils. The next question, posed within the framework of risk control, is to determine the consequences of such a sand-carrying well if it does develop.

Required thickness of the covering layer

Different authors determine the required thickness of a covering sand layer by analysing the stability of a sand prop in a hole in a grout layer (Bieberstein [6], Sellmeijer [7]). Sellmeijer considers the importance of a surcharge at the top of the sand prop. By lack of effective stress at the top the prop looses its stability, initiating a sand boil. Tol et al. [8] did experiments to establish the friction of the sand in the hole. The experiments resulted in values of *Ktgd* as a function of the porosity of the sand. Two branches appeared, one for medium to loose sands with *Ktgd* between 0.4 and 0.5 and one for dense sands with *Ktgd* between 1.0 to 1.2. This corresponds to an active and passive condition in the hole at failure along a vertical wall.

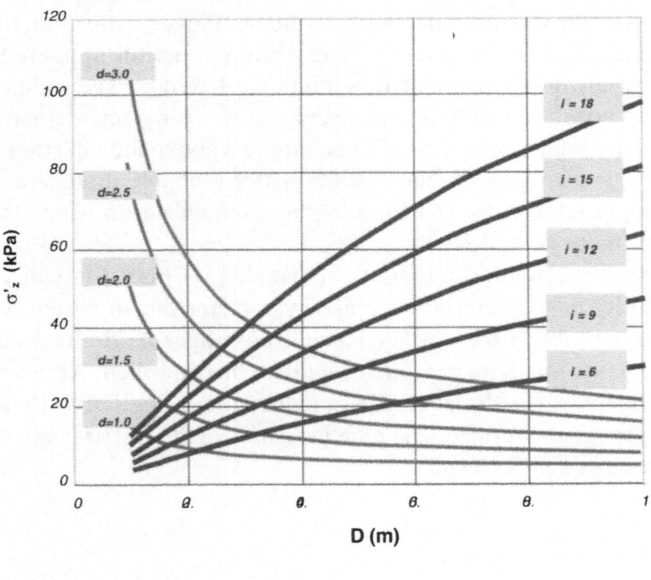

Figure 2. Required thickness d of a covering sand layer as function of gradient i_d.

Figure 2 (Tol et al. [8] presents the required thickness of a covering sand layer. The overall safety factor η =2 is applied to the height of the covering layer. For the parameters design values are used: $K_d tg\delta = 0.32$, $\gamma_{sat} = 18.2$, and i as shown in Figure 2. The thickness of the groutlayer is equal to $L_d = 1.25$ and in the friction in the covering layer is calculated with K_r tg φ equal to 0.125. It should be noted that the design method presented is valid for homogeneous soil in the hole and the covering layer.

Scouring under a grout layer

For every design, it is recommended that the consequences of any failure in design or construction be examined, together with the scenarios one can rely on. Within this context, the consequences of a sand-carrying well through the grout arch in the Tram Tunnel Project in The Hague were recently considered. The starting point is that a sand boil develops. The question to be answered is how much soil is transported from under the grout arch into the pit and over what time period, and is the scouring limited to underneath the construction pit or does it extend outside the retaining walls affecting foundations. This question is extremely relevant as the answer indicates how much time is available to take measures to stop the flow.

Delft Hydraulics has carried out analyses to find the answer (Mastbergen et al. [9]). The approach consists of linking several different processes, which are described in analytical models. The problem is dominated by the discharge conditions below the arch. Before scouring occurs, the (permeability of the) sand beneath the arch and the diameter of the hole determine the discharge of the well for a given pressure difference. Scouring causes the discharge to increase rapidly, because it is now the scour area that determines the discharge. Delft Hydraulics considered that scouring causes a disturbance under the grout arch, which extends under the arch by regressive erosion (Figure 3).

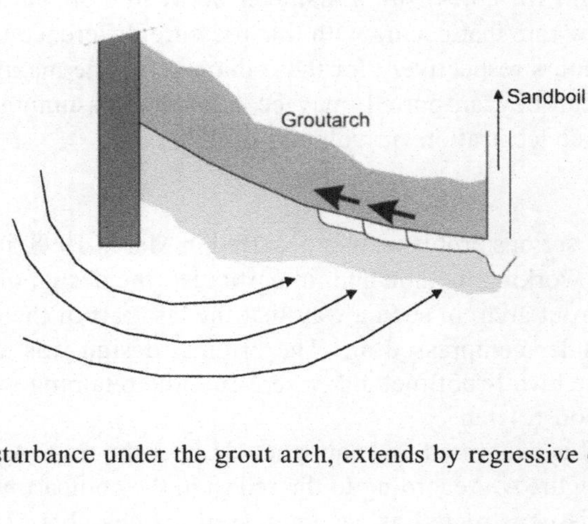

Figure 3. Disturbance under the grout arch, extends by regressive erosion.

This phenomenon shows a strong analogy with the advance of a breach along an underwater slope during sand extraction and is familiar from dredging techniques. The increase of the scour surface is thought to be equal to the speed of the breach, described with the equation (Rhee and Bezuijen [10])

$$v_b = -(i + k_l\ (1-n_0)R\sin(\varphi-\alpha)/\sin\varphi)/\ \Delta\ n$$

where:

- v_b = velocity of a breach (local failure) in m/s
- n_0 = porosity of the sand
- i = gradient (negative in the case of an outflow)
- R = relative density of the sand: $R = (\gamma_{solids}-\gamma_w)/\gamma_w$ = 1,65
- Δn = increment of the sand porosity from in-situ density to loose
- k_l = permeability of loose sand
- α = slope of the failure (for small failure height almost 90°)
- φ = friction angle of the in-situ sand

This equation can be used to determine development of the scouring surface under the arch. This surface area determines the flow rate. A groundwater flow calculation can be used to determine the hydraulic gradient at the erosion surface. The hydraulic gradient however forms the input for the above formula. The calculation is therefore carried out iteratively. If scouring continues, it appears that the flow rate initially increases substantially until there is ultimately an almost constant flow rate. Because of the decreased hydraulic gradient an equilibrium profile is in fact created under the grout arch. The maximum flow rate is strongly influenced by the permeability of the soil under the grout arch.

Delft Hydraulics also examined whether the discharge capacity of a hole limits the flow rate of a sand-water mixture, and subsequently the scouring. It seems that with a pressure difference of approximately 9 m water column such as found at the Tram Tunnel, the discharge capacity of the holes does not form a limit for holes with a diameter between 0.08 and 0.12 m. The maximum flow rate that occurs with this pressure difference is between 70 and 180 m3/hours respectively for these diameters. The maximum volume of sand that can be transported, may be calculated assuming a maximum average sand concentration (in volume) of 30%.

Evaluation

Following the serious problems that occurred in March 1998, it was decided to change the working method and, in particular, the design of the building pit with the grout arch, in such a way that the last part of the excavation is carried out under compressed air. The original design was already based on the so-called wall-roof method, where first the retaining walls and then the roof are constructed.

The tunnel is excavated and constructed under the roof. A cross-section is shown in Figure 4. According to the redesign the compartment under the lowest floor is constructed as a compressed air chamber. To reduce the enormous costs of the work under compressed air, it was decided to excavate 3.25 m under the lowest floor at atmospheric conditions. In this situation, the stability of the sand prop in expected holes in the grout arch is achieved based on the above-mentioned analyses. Before the start of excavation, a drainage system is installed with wells close to the walls.

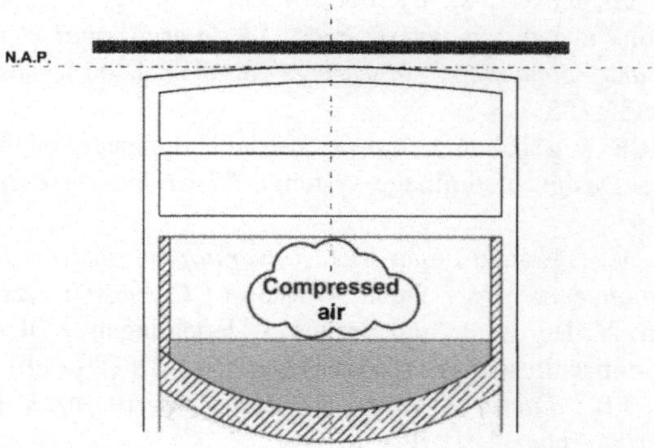

Figure 4. Cross-section of the Tramtunnel with the chamber for compressed air.

During the atmospheric excavation, several boils developed that were controlled and repaired under compressed air by additional grouting. According to the analyses based on homogeneous soil, however, such boils could not occur in this stage. In fact, the soil just above and at the level of the grout arch is not uniform. Several thin clay layers are present in the covering layer. This could explain why actual conditions were worse then analysed. If a clay layer is present in the covering sand layer and the drainage system is installed above this clay layer, then excess pore pressure will build up between the grout layer and the clay layer if there are holes in the grout layer. If the hydraulic resistance of the clay layer considerably exceeds the resistance of the grout layer, then the excess pore pressure may rise to such a level that the effective stresses at the top of the hole in the grout layer are too small and the sand prop in the hole looses its stability.

The conclusion regarding the stability of sand props in a jetgrout layer is that a drainage system should be installed as deep as possible, and at least below the lowest silt and clay layers.

Regarding scouring under the grout arch, some sand flow into the pit before the boils were controlled. At one location grouting was applied to fill the space under the grout arch formed by the scour. As well the discharge of the water sand mixture trough this boil as the amount of grout needed to fill the gap was regarding the order of magnitude in accordance with the above presented approach of Delft Hydraulics.

References

1. Tol, A.F. van : Geotechnical aspects of the Willems-Railway-tunnel, Jetgrouting as an alternative for groundwater retaining. *Civiele Techniek*, 46/3 (1991), pp.16-25.
2. Tol, A.F. van: Panelgrouting in building pit Binnen Rotte, A case history with a high degree of complexity. *Civiele Techniek*, 54/3, (1999), pp.22-25.
3. Tol, A.F. van, Koster, S., Ramler, J.P.G., Vrijling, J.K., and Verruijt, A.: Imperfections in jetgrout layers, *Proc. XV International Conference on Soil Mechanics and Foundation Engineering* (Togrol et al., ed.). Balkema, Rotterdam, 2001, pp.1883-1886.
4. Elprama, R.E. and Tol, A.F. van: Geotechnical aspects of the Willems-Railway-tunnel, Design of drainage systems. *PT/Civiele Techniek*, 43/3, (1988), pp.35-41.
5. Hoek, van der E.E., and Luger, H.J.:*Putverstopping Station Spui, Causes and possible counter measures,* Geodelft Report CO-397880 vers.2, 2001.
6. Bieberstein, A., Herbst, J., and Brauns, J.: Hochliegende Dichtungssohlen bei Baugruben-umschliessungen. *Geotechniek*, Vol 22(2), (1999), pp.114-123.
7. Sellmeijer, J.B.: *Safe excavation level above jetgrout arch.* GeoDelft Report CO-390471/8, Annex 1, 1999.
8. Tol, A.F. van, Riel, A.J.E. van, Vrijling, J.K.: Towards a reliable design for jetgrout layers. *Proc. XV International Conference on Soil Mechanics and Foundation Engineering* (Togrol et al., ed.). Balkema, Rotterdam, 2001, pp.1887-1890.
9. Mastbergen, .R. and Kesteren, W.G.M. van: *Scour under a grout arch, Risque analyses of scour processes at the Tramtunnel project in The Hague*, Delft Hydraulics Report, 2001.
10. Rhee, C. van, and Bezuijen, A.: Influence of seepage on stability of sandy slope. *J. of Geotechnical Eng.*, Vol. 118/8 (1992).

The Flushing Effect

GERARD J.M. UFFINK
University of Technology, Delft & National Institute of Public Health and the Environment, Bilthoven, The Netherlands

A common situation in coastal aquifers is to find fresh water flowing over a more or less stagnant body of saltwater. Since salt and fresh water behave as miscible fluids, the interface between the two fluids is expected to become more and more diffuse, but in practice the interfaces remain relatively sharp. It is believed that the continuous flow of fresh water is responsible for this phenomenon. The fresh water removes the salt and keeps the transition zone small. By intuition it is felt that the magnitude of the fresh water velocity also plays a role, i.e. the higher the fresh water velocity, the sharper the interface. This supposed influence of the fresh water velocity is called the flushing effect. Verruijt [1] was the first to address this subject in a paper entitled 'Steady dispersion across an interface in a porous medium'.

Verruijt's solution

Verruijt uses two different models to describe the situation. In the first both fluids (fresh and salt groundwater) have the same horizontal velocity u. This velocity is constant throughout the domain. The vertical velocity v is zero everywhere. For the steady state the following differential equation holds:

$$u\frac{\partial c}{\partial x} = D_L \frac{\partial^2 c}{\partial x^2} + D_T \frac{\partial^2 c}{\partial y^2} \tag{1}$$

c is the salt concentration, D_L and D_T are the longitudinal and transverse dispersion and u is the groundwater velocity in x-direction. For the dispersion coefficients the relations $D_L = \alpha_L |u|$ and $D_T = \alpha_T |u|$ are used, where α_L and α_T are the longitudinal and transverse dispersivities. The original interface (i.e. before mixing) is located along the positive x-axis. Mixing starts at $x =$ 0, Verruijt gives the full solution as an integral expression, which for large values of x (large compared to α_T) may be approximated by:

$$\frac{c}{c_s} = \frac{1}{2} erfc\left(\frac{y}{2\sqrt{\alpha_T x}}\right) \tag{2}$$

c_s is the concentration in the saline water. The solution only depends on α_T. Apparently, it is irrelevant whether fresh water flows slowly or fast.

In the second model the salt water is stagnant and fresh water flows at a constant velocity u. Between the fluids a resistance layer is assumed that controls the movement of salt across the interface by the following condition:

$$x > 0, y = 0;\; D_T \frac{\partial c}{\partial y} = \frac{c - c_s}{h} \tag{3}$$

The concentration in the salt water is assumed to remain constant. The parameter h is a measure for the resistance to the flux of salt. The full solution is given again as an integral expression. For large x it reduces to:

$$\frac{c}{c_s} = erfc\left(\frac{y}{2\sqrt{\alpha_T x}}\right) - \frac{h\alpha_T u}{2\sqrt{\alpha_T x}} \exp\left(-\frac{y^2}{4\alpha_T x}\right) \quad for\ y > 0 \tag{4}$$

The influence of the velocity u is clearly seen. A large value of u decreases the concentration, which confirms the flushing effect. However, a comment on the resistance parameter h should be made. Verruijt refers for this concept to the theory of heat conduction. In Carslaw and Jaeger [2] (page 19-20) a similar resistance layer is applied for the heat flux to a cooling liquid flowing along a warm body. If the analogy with heat flow is correct, then the coefficient h is a function of u. Carslaw and Jaeger report for turbulent flow over a heated pipe that h is proportional to $u^{-.8}$. This is of particular interest. With h proportional to u^{-1}, the velocity cancels out, while for any negative power other than -1 flushing still occurs, but significantly less than (4) suggests. So, the question remains: is there a flushing effect and if so, how strong is it.

Numerical Simulation

The problem is further analyzed by a random walk technique. In the random walk hypothetical elements (particles) are introduced that usually carry an amount of solute mass. We use a different approach here: particles that carry a certain amount of vorticity (random vortices). The reason to do so is explained below. First, a related problem described by De Josselin de Jong and Van Duijn [3] is discussed. This problem is used also for validation of the proposed approach with random-vortices. De Josselin de Jong and Van Duijn solved a one-dimensional case given by the following equation:

$$\frac{\partial c}{\partial t} = \frac{\partial}{\partial y}\left(D_T \frac{\partial c}{\partial y}\right) \tag{5}$$

Note that D_T ($D_T = \alpha_T |u|$) is not a constant, but a function of y and t. Initially the interface is sharp and positioned at an angle β with the horizontal. The y-direction is perpendicular to the interface. The effect of density difference is taken into account. Elaboration of the equilibrium condition yields:

$$u-u_s=-\frac{\kappa g}{\varepsilon\mu}(c-c_s)\sin\beta \tag{6}$$

κ is the intrinsic permeability of the aquifer, ε is the porosity, μ is the dynamic viscosity of the groundwater and g is the constant of gravity. Initially, an abrupt change in concentration occurs at the interface and eq. (6) reduces to the well-known Edelman [4] equation for shear flow. De Josselin de Jong and Van Duijn considered several situations with salt and fresh water flowing at different velocities. Here, we look at the case where salt water is stagnant. Then, for $y \rightarrow \infty$ we can write $c = 0$ and $u = u_f$, while for $y \rightarrow -\infty$ $c = c_s$ and $u_s = 0$. Accordingly, equation (6) is rewritten to:

$$\frac{u}{u_f}=1-\frac{c}{c_s} \tag{7}$$

In equation (5) c can now be replaced by u, which has some benefits. The random walk requires that.(5) is rewritten to a Fokker-Planck equation, e.g.:

$$\frac{\partial c}{\partial t}+\frac{\partial}{\partial y}\left(c\frac{\partial D_T}{\partial y}\right)=\frac{\partial^2}{\partial y^2}(D_T c) \tag{8}$$

The right hand side corresponds to the second derivative term in the Fokker-Planck equation. However, the rewriting brings along a term at the left hand side that looks like an 'advection' term with a 'velocity' $\partial D_T/\partial y$. Such a complication does not occur when first c is replaced by u, the entire equation is differentiated to y and finally in (5) the vorticity ω is introduced:

$$\omega=\frac{\partial u}{\partial y} \tag{9}$$

Then, the governing equation becomes:

$$\frac{\partial \omega}{\partial t}=\frac{\partial^2}{\partial y^2}(D_T\omega) \tag{10}$$

Equation (10) is a Fokker-Planck equation for system of particles carrying an amount of vorticity. Thus, we may simulate the problem with N random vortices. The total vorticity in the system is $\Delta u = u_f - u_s = u_f$. Each vortex carries a vorticity $\Delta u/N$. One complication remains: the (random) displacement depends on D_T, which depends on u, which in turn depends on the position of the vortices. Therefore, the distribution of the velocity must be evaluated each time-step.

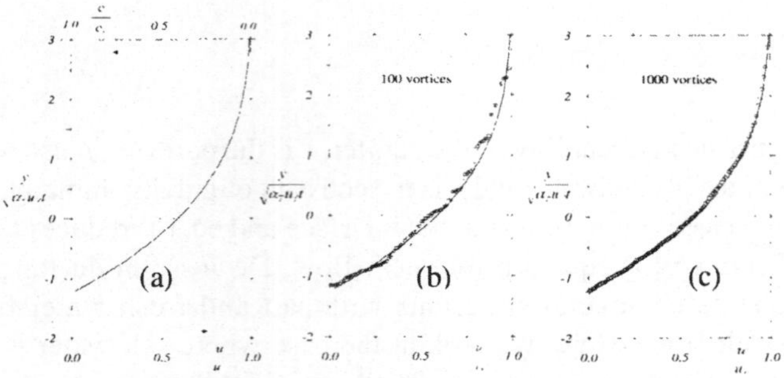

Figure 1. Velocity and concentration distribution in the transition zone. (a) Solution by Josselin de Jong and Van Duijn (separatrix). (b) and (c) comparison of simulation results with separatrix.

Simulations were carried out and compared with De Josselin de Jong's solution. The curve by De Josselin de Jong and Van Duijn (Fig. 1a) is referred to as the separatrix. The parameter at the vertical axis includes the fresh water velocity u_f. This means that a higher velocity accelerates the mixing process, but the concentration at $y = 0$ does not alter. Thus, flushing does not occur. Figure 1b shows simulation results for 100 vortices. The fit is relatively good for this small number of vortices. For 1000 vortices (Figure1c) an excellent fit is obtained. Here only every 10th particle is plotted for reasons of clarity.

Two-dimensional case

In 2-D eqation. (7) reappears as a simplification of the vorticity equation:

$$\frac{\partial u}{\partial y} - \frac{\partial v}{\partial x} = -\frac{\kappa g}{\varepsilon \mu}\left(\frac{\partial c}{\partial y}\sin\beta - \frac{\partial c}{\partial x}\cos\beta\right) \tag{11}$$

Uffink [5] uses a boundary layer approximation to simplify (11). This approach requires that x is large compared to α_T, which is the same condition used by Verruijt to obtain the expressions (2) and (4). In the boundary layer approach (11) reduces to:

$$\frac{\partial u}{\partial y} = -\frac{\kappa g}{\varepsilon \mu}\frac{\partial c}{\partial y}\sin\beta \tag{12}$$

After integration, equation (6) and (7) are obtained. A steady state solution can now be found by a similarity transformation. Figures (2a) and (2b) give the distribution of the velocity components u and v. Again, the solution depends only on the geometry and α_T, so the flushing effect does not appear.

Since (7) holds for the 2-D case, the basic equation can be written in terms of the vorticity, which allows a simulation with random vortices. Results are compared with the separatrix and the steady state (Figures 3 and 4).

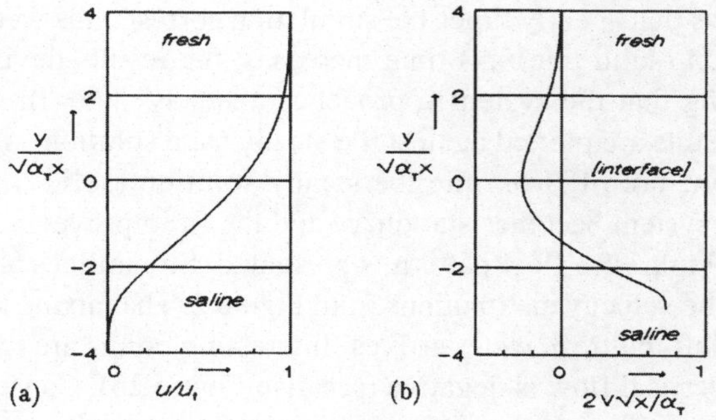

Figure 2. Distribution of the velocity components (a) parallel to the interface and (b) perpendicular to the interface.

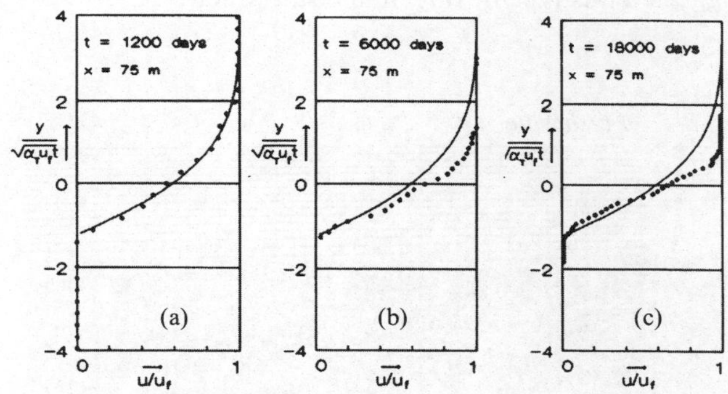

Figure 3. Simulation results compared with the non-steady state solution by De Josselin de Jong and Van Duijn.

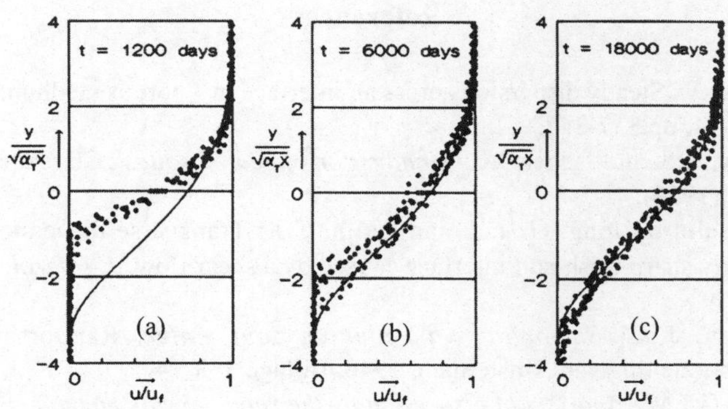

Figure 4. Simulation results compared with the 2-D steady state solution by Uffink.

Figure 3 shows that at early times the simulation corresponds well to the non-steady solution (solid line). As time increases, the results deviate from the line, indicating that the system approaches a steady state. In Figure 4 the simulation results are plotted against the steady state solution. At early times the simulations are off from the the steady solution (solid line). As time increases the system becomes stationary and the fit improves.

Finally in Figure 5 a flow pattern is presented. Streamlines are calculated according to the velocity distributions from Figure 2. The mixing starts at point O. From the left the fresh water arrives. Interesting points are the following. At $y = 0$ the vertical flow is negative (see also Figure 2b). Consequently, the streamline starting from O dips into the zone $y < 0$ (the red dashed line denotes the x-axis). Figure 2b also shows a positive vertical flow for large negative values of y. This means in the saline water an upward flow exists, that replenishes the salt that is 'washed away' by the fresh water. Cooper [6] already suggested this type of flow in the salt water.

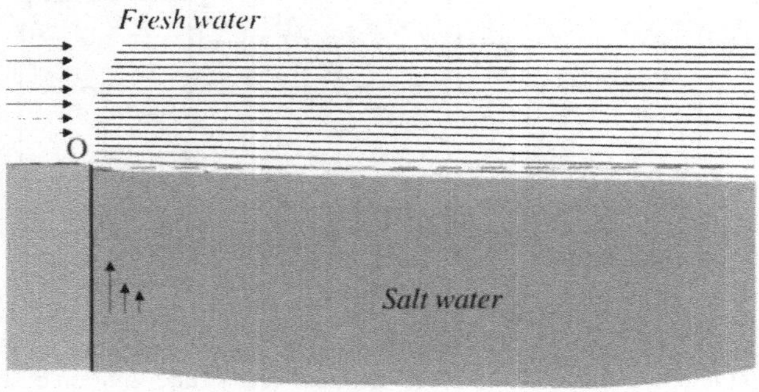

Figure 5. Flow pattern in salt water and fresh water for steady state.

References

1. Verruijt, A.: Steady dispersion across an interface in a porous medium. *J. Hydrol.* (1971), 14, pp.337-347.
2. Carslaw, H.S. and Jaeger, J.C.: *Conduction of heat in solids* , Clarendon, Oxford, 2nd ed.,(1959).
3. De Josselin de Jong, G. and Van Duijn, C.J.: Transverse dispersion from an originally sharp fresh-salt interface caused by a shear flow. *J. Hydrol.* 84 (1986), pp. 55-79.
4. Edelman, J.H.: *Strooming van zoet en zout water*. Rapport inzake de watervoorziening van Amsterdam, 1940, Bijlage 2: 8-14.
5. Uffink, G.J.M.: *Analysis of dispersion by the random walk method.* Thesis Delft University of Technology, (1990).
6. Cooper, H.H.: A hypothesis concerning the dynamic balance of fresh and salt water in a coastal aquifer. *J. Geophys. Res. 64* (4), (1959), pp. 461-46.

CONSOLIDATION & CREEP

The Squeeze Problem

FRANS B.J. BARENDS
GeoDelft & TUDelft, Delft, the Netherlands

Instantaneous pore pressures after sudden uniform loading are, according to standard soil mechanics, equal to the loading applied. This is the outcome of the oedometer test. In case of compressible pore water or grains a reduction factor is due, but the behaviour of the immediate reaction is a rather uniform pore pressure distribution. However, the instantaneous pore pressure development may be completely different than observed in the oedometer test. This is caused by special constraints, different from the oedometer. Such a situation is squeeze consolidation.

It was noticed by Mandel [1] in a brilliant exposé extending Terzaghi's view on consolidation. The axisymmetric squeeze problem was introduced by Rendulic [2], Barron [3] and Kjellman [4] with regard to the application of vertical drains. Their solution is widely used. At regular times the subject receives new attention, see Richart [5], de Leeuw [6], Hansbo [7], Gibson [8], and Abousleiman [9].

In this article a fresh look to the squeeze problem provides a surprising result. Moreover, an analytic solution is obtained valid for generally nonlinear soil behaviour. It shows a pronounced effect of load history on the consolidation process and this sheds new light on the consolidation factor and the evaluation of the operation of vertical drains.

Verruijt's simple method for 3D consolidation

Verruijt [10] posed an elegant and simple approach for 3D consolidation problems, suitable for qualitative judgement. His solution for a space-average pore pressure P, assuming a parabolic distribution, yields

$$P = P_0 \exp[-t/t_0] \quad \text{with} \quad t_0 = L_1 L_2/(4c) \tag{1}$$

The hydraulic period $T_{99\%}$ is therefore achieved at $4.6t_0 = 1.15L_1L_2/c$. Notice the distinction between the two length scales. L_1 is the average flow path, connecting the location of maximum pressure to the drained border, and L_2 is the hydraulic radius, related to the dissipation capacity (volume over drained surface). For 2D consolidation $t_0 = L_1L_2/(3c)$. The determination of L_1 and L_2 is crucial. It requires some experience. Verruijt's method represents a powerful tool for quick estimation of the general behaviour of the consolidation process in any specific situation, while taking care of boundary and initial conditions in a proper way. Verruit's assumption of the geometric parabolic shape is, by chance, exact in a general squeeze problem, also for non-linear soil behaviour.

The squeeze problem

Consider a plane-symmetric situation. A saturated soil sample, free at the sides, is squeezed by a line load *2F* between two smooth and rigid plates. The pore water is assumed incompressible and excess pore pressures dissipates in horizontal direction. The soil behaves linearly elastic and the soil condition is such that no horizontal deformation occurs and horizontal stresses are disregarded, i.e. Poisson's constant $\nu = 0$. Thus, only vertical stresses and deformations are considered. Therefore, the situation complies with the following field variable dependencies

$$\sigma_v = \sigma[x]; \quad F = \int_0^L \sigma\, dx; \quad \sigma_v{}' = \sigma'[t], \; \varepsilon_v = \varepsilon\,[t]\;; \quad p[x,t] = \sigma[x] - \sigma'[t]$$

The rigid plate allows stress redistribution in the soil sample, so that the boundary conditions at the sides are satisfied at all time. These conditions are (only $0<x<L$ is considered) at $x = 0$: $q = 0$, $p_{,x} = 0$, $\sigma_{,x} = 0$ and at $x = L$: $p = 0$. The volume compressed in *dxdt* is $dV_c = h_{,t}dxdt$. Vertical strain is uniform, not dependent on x, but dependent on time. Strain increment is related to the change in height, according to $d\varepsilon = dh/h$. Thus, $dV_c = h_{,t}dxdt = h\varepsilon_{,t}dxdt$. Linear-elastic soil behaviour is expressed by the constitutional law $\sigma' = -\varepsilon/\alpha$, where α is the laterally confined compressibility (for $\nu = 0$). The minus sign reflects that strain is positive for extension and stress for compression. Note, σ' is independent of x. Thus, $dV_c = -\alpha h\sigma'_{,t}dxdt$. The volume compaction is compensated by the average pore water volume outflow dq in time period dt. Thus: $dV_o = -q_{,x}hdxdt$. The minus sign corresponds to the definition of positive volume increase and positive outflow.

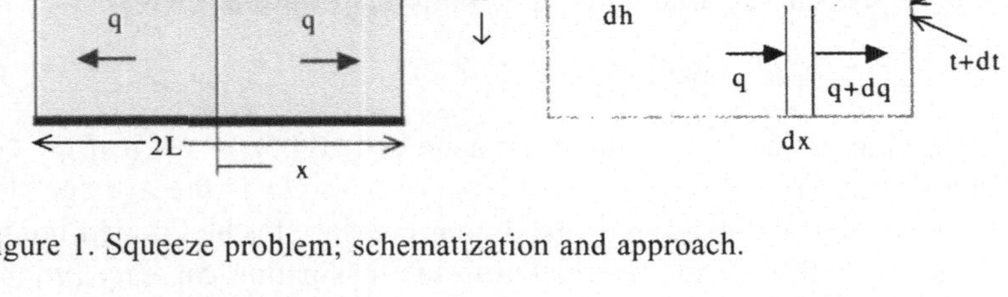

Figure 1. Squeeze problem; schematization and approach.

Mass balance requires $dV_c = dV_o$ or $\alpha\sigma'_{,t} = q_{,x}$. Introduction of Darcy's law $q_{,x} = -(k/\gamma)\, p_{,x}$ yields $\sigma'_{,t} = -c\, p_{,xx}$ with $c = k/\gamma\alpha$. Since σ' is only dependent on time, this equation can be integrated straight forwardly, which yields $cp_{,x} = -\sigma'_{,t}\, x + C_1$. The boundary condition at $x = 0$ gives $C_1 = 0$. Next, $cp = -\frac{1}{2}\sigma'_{,t}x^2 + C_2$ and the boundary condition at $x = L$ gives $C_2 = \frac{1}{2}L^2\sigma'_{,t}$. Thus

$$cp = \tfrac{1}{2}\sigma'_{,t}\,(L^2-x^2) \tag{2}$$

Hence, the pore pressure distribution is parabolic in space at any time with a maximum $L^2\sigma'_{,t}/2c$ that depends on time. If the loading is sudden, the response is an immediate parabolic pore pressure in space and, accordingly, a sudden flow, increasing from the centre to the sides. This situation can be expressed in terms of the loading F, as follows

$$F = \int_0^L \sigma\, dx = \int_0^L (\sigma' + (1/2c)\sigma'_{,t}(L^2-x^2))\, dx = \sigma' L + (L^3/3c)\sigma'_{,t} \tag{3}$$

It represents a normal differential equation in time in terms of the effective stress σ'. The method of separation of variables is applied for its solution. Note, σ' and F are only functions of t. Separating the differentials yields $(3c/L^2)d[t] = (F/L-\sigma')^{-1}d[\sigma']$. For the loading condition $F = 0,\ t < 0$ and $F = F_1 = constant,\ t > 0$, we integrate for $t > 0$, according to

$$(3c/L^2)\, d[t] = (F_1/L - \sigma')^{-1}\, d[\sigma'] = -\, d[\ln[F_1/L - \sigma']]$$

which yields with $\sigma' = \sigma'_0 = 0$ at $t = 0^+$

$$(3c/L^2)\int_0^t dt = 3ct/L^2 = -\int_0^{\sigma'} d[\ln[F_1/L - \sigma']] = -\ln[(F_1/L - \sigma')/(F_1/L)]$$

From which follows $\sigma' = (F_1/L)(1 - \exp[-3ct/L^2])$. Consequently, we obtain $\sigma'_{,t} = (3cF_1/L^3)\exp[-3ct/L^2]$ and with equation (2) the pressure becomes

$$p = (3/2)\,(F_1/L)\,(1-(x/L)^2)\,\exp[-3ct/L^2] \tag{4}$$

This solution shows that a parabolic pore pressure arises with a maximum in the centre of the sample, immediately. At $t = 0^+$ it equals 1.5 times the average loading F_1/L. The pore pressure dissipation reaches 99% completion for $t = 1.54L^2/c$ (hydrodynamic period). Subsequently, the compaction of the sample proceeds according to

$$S = -\int_0^h \varepsilon\, dy = -\,\varepsilon\, h = \alpha\sigma' h = (\alpha F_1 h/L)(1 - \exp[-3ct/L^2]) \tag{5}$$

This result is in character similar to the oedometer test (vertical consolidation). The factor L^2 represents the hydraulic radius (drainage capacity) and seepage length for squeeze condition. The vertical compressibility is embodied in coefficient c, just like in the oedometer test, except here in the absence of horizontal deformation ($\nu = 0$). The approach can be extended to a general non-linear soil behaviour, with variable compressibility and permeability. Surprisingly, at any time the instantaneous pore pressure will be parabolic in space as well.

Non-linear squeeze consolidation

A realistic behaviour of the soil compaction is the so-called hyperbolic law. It is expressed in terms of absolute stress: $\sigma' = \sigma'_1+\sigma'_0$, where σ'_0 is the initial effective stress at strain $\varepsilon = \varepsilon_0$, and σ' is the stress at corresponding strain $\varepsilon = \varepsilon_1+\varepsilon_0$; σ'_1 and ε_1 are incremental values. The hyperbolic law states $d[\varepsilon] = -(\eta/\sigma')d[\sigma']$ or $\varepsilon - \varepsilon_0 = -\eta \ln[1+\sigma'_1/\sigma'_0]$ or $\exp[\varepsilon_1] = (1+\sigma'_1/\sigma'_0)^{-\eta}$. Characteristic values for the so-called compression index η are 0.8 (mud), 0.1 (peat), 0.02 (silt/clay), and 0.005 (sand).

Also the permeability effect can be incorporated, stating that as compaction proceeds the permeability decreases, according to $d[k]/k = \chi d[\varepsilon]$ or $k = k_0 exp[\chi(\varepsilon-\varepsilon_0)]$ or $k = k_0(1+\sigma' 1/\sigma'_0)^{-\chi\eta}$. For practical cases $0 \leq \chi < 2$.

Because for the squeeze situation, considered here, ε is only a function of t, also k is only a function of t. The permeability effect generates an extra term in the corresponding storage equation, but it vanishes, as $k_{,x} = 0$. Hence

$$(\eta/\sigma')\sigma'_{,t} = -((k/\gamma)\, p_{,x})_{,x} = -(k/\gamma)\, p_{,xx} = -(k_0/\gamma)\, (1+\sigma'_1/\sigma'_0)^{-\chi\eta}\, p_{,xx}$$

Thus the storage equation is

$$\sigma'_{,t} = -\, c_2\, p_{,xx} \qquad (6)$$

with $c_0 = k_0\sigma'_0/\gamma\eta$, $\theta = 1-\chi\eta$, and $c_2 = (k_0/\gamma\eta)\,(1+\sigma'_1/\sigma'_0)^{-\chi\eta}(\sigma'_1+\sigma'_0) = c_0\,(1+\sigma'_1/\sigma'_0)^{\,\theta}$. The practical value of the $\chi\eta$ is normally somewhere between 0 and 2. For $\theta = 0$ the storage equation is linear, as the non-linearity in permeability and compressibility effects cancel out for the pore pressure generation; soil behaviour is, however, yet non-linear. Because σ' and c_2 are not functions of x, spatial integration of equation (6) can be performed. It shows that during consolidation the pore pressure distribution in space remains parabolic, with a maximum dependent on time.

$$p_{max} = \sigma'_{,t}L^2/2c_2 \qquad (7)$$

The non-linear total equilibrium equation becomes $F = F_1+F_0 = (\sigma'_1+\sigma'_0)L + (L^3/(3c_2))\sigma'_{,t}$ or $(L^2/(3c_2))\sigma'_{,t} = F_1/L-\sigma'_1$, where F_0 is the initial state loading until $t = 0$ (consolidation completed: $p = 0$) and F_1 is the constant additional loading step from $t = 0$ onward, inducing pore pressures (consolidation). Use has been made of the equilibrium state for $t < 0$, in fact $F_0/L = \sigma'_0$. Elaboration shows that the maximum pressure is expressed by

$$p_{max} = L^2\sigma'_{,t}/2c_2 = (3/2)\,(F_1/L-\sigma'_1) = (3/2)\,(F/L-\sigma') \qquad (8)$$

Since for incompressible pore water (and $v = 0$) $\sigma'_1 = 0$ at $t = 0^+$, this result shows that the instantaneous maximum pore pressure is 1.5 times the average loading F_1/L, similar as found for the linear case. Furthermore, an analytic solution can be obtained for the total equilibrium equation, using separation of variables. Assuming initial and final conditions, according to

$$t < 0 \quad F = F_0 + F_1,\ \sigma'_0 = F_0/L,\ F_1 = 0\ ,\ \varepsilon = \varepsilon_0 = \sigma'_1 = p = 0 \tag{9a}$$
$$t = 0^+ \quad F = F_0 + F_1,\ F_1 \neq 0,\ \varepsilon_0 = \sigma'_1 = 0,\ p \neq 0 \tag{9b}$$
$$t > 0 \quad F = F_0 + F_1\ (F_1 = constant) \tag{9c}$$
$$t \rightarrow \infty \quad F_1 = \sigma'_1 L,\ p = 0 \tag{9d}$$

the total equilibrium equation is reformulated for $t > 0$, which yields

$$\int_{a-\xi}^{a} \psi^{-1} (b - \psi)^{-\theta} d\psi = 3\tau \tag{10}$$

with $b = 1 + a$, $a = F_1/F_0$, $\xi = \sigma'_1/\sigma'_0$, $\theta = 1 - \chi\eta$, $\tau = t/t_0$, $t_0 = L^2/c_0$. Further integration is in principle possible for general values of θ, but it leads to complicated series. We can develop simple solutions for some values of θ. Elaboration for the effective total stress σ', the maximum pore pressure p_{max} and the settlement $S_1 = -\varepsilon_1 h = \eta \ln[1 + \sigma'_1/\sigma'_0]$ yields

$$\theta = 0 \quad \sigma' = F/L - (F_1/L)\exp[-3c_0 t/L^2] \tag{11a}$$
$$p_{max} = (3/2)(F_1/L)\exp[-3c_0 t/L^2]$$
$$S_1 = \eta h \ln[1 + (F_1/F_0)\exp[-3c_0 t/L^2]]$$

$$\theta = ½ \quad \sigma' = (F/L)((1-E)/(1+E))^2 \tag{11b}$$
$$\text{with } E = (\sqrt{[F]} - \sqrt{[F_0]})(\sqrt{[F]} + \sqrt{[F_0]})^{-1}\exp[-3c_{½} t/L^2]$$
$$p_{max} = (3/2)(F/L)(4E)(1 + E)^{-2}$$
$$S_1 = \eta h \ln[(1 + (F_1/F_0)((1-E)(1+E))^2]$$

$$\theta = 1 \quad \sigma' = (F/L)(1 + (F_1/F_0)\exp[-3c_1 t/L^2])^{-1} \tag{11c}$$
$$p_{max} = (3/2)(F/L)(1 + (F_0/F_1)\exp[+3c_1 t/L^2])^{-1}$$
$$S_1 = -\eta h \ln[1 - (F_1/F)(1 - \exp[-3c_1 t/L^2])]$$

whith $c_0 = k_0 F_0/(\gamma\eta L)$, $c_{½} = k_0\sqrt{[FF_0]}/(\gamma\eta L)$, and $c_1 = k_0 F/(\gamma\eta L)$. In all situations the initial maximum pore pressure is equal to $(3/2)(F_1/L)$ as can be checked easily. Note, for $t = 0^+$, $((1-E)/(1+E))^2 = F_0/F$.

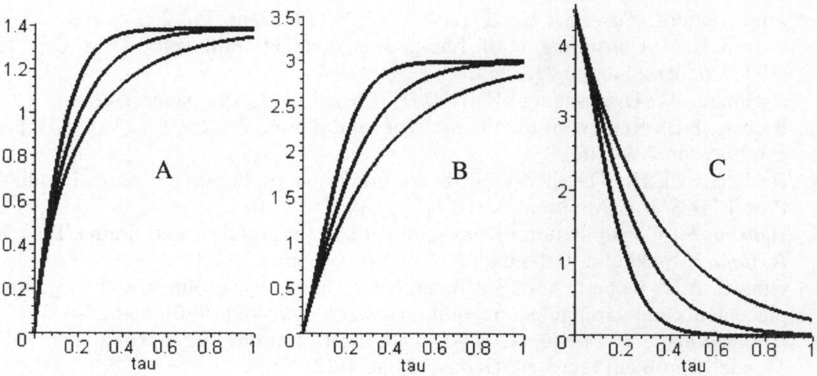

Figure 2. Stress σ'_1/σ'_0 (A), settlement $S_1/(\eta h)$ (B), and pore pressure p/σ'_0 (C) for load factor 3; blue $\theta = 1$, green $\theta = ½$, and red $\theta = 0$.

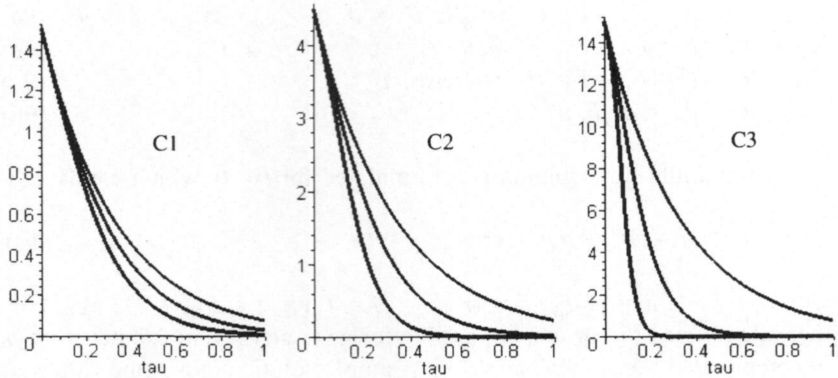

Figure 3. Pore pressure p/σ'_0 for load factor 1 (C1), 3 (C2), and 10 (C3); blue $\theta = 1$, green $\theta = ½$, and red $\theta = 0$.

In Figure 2A the effective stress σ'_1/σ'_0, in 2B the settlement $S_1/(\eta h)$, and in 2C the pore pressure p/σ'_0 are shown against time $\tau = t/t_0$ for the situation $F_1/F_0 = 3$, and for $\chi\eta = 0$, ½, and 1 (or $\theta = 1$, ½, 0). The effect of the non-linearity shows that the consolidation period (hydrodynamic period) is strongly reduced, up to a factor 5 between $\chi\eta$ is 0 and 1.

In Figure 3 the pore pressure p/σ'_0 is plotted against time $\tau = t/t_0$ for different load factors; C1: $F_1 = F_0$, C2: $F_1 = 3F_0$, and C3: $F_1 = 10F_0$. These results show that the soil behaviour in the squeeze test strongly depends on load history. Although this observation is proven valid for the particular case $\nu = 0$, it is believed that this tendency is true for general cases.

References

1. Mandel, J.: Consolidation des sols (etude mathématique). *Geotechnique* 3(7) (1953), pp.287-299.
2. Rendulic, L.: Der hydrodynamische Spannungsausgleich in zentral entwässerten Tonzylindern. *Wasserwirtsc. & Technik* 23/24 (1935), pp.250-253 & 269-273.
3. Barron, R.A.: Consolidation of fine-grained soils by drain wells. *Tr. ASCE* 113 (1947), paper 2246, pp.718-754.
4. Kjellman, W.: Discussion to Barron. *Tr. ASCE* 133 (1948), paper 2340.
5. Richert, F.E.: Review of the theories for sand drains. *Tr. ASCE* 124 (1957), paper 2999, pp.709-736.
6. De Leeuw, E.H.: The theory of 3D consolidation applied to cylindrical bodies. Proc *VI ICSMFE*, Montreal, Vol 3 (1965), pp.287-290.
7. Hansbo, S.: Consolidation of fine-grained soil by prefabricated drains, Proc X *ICSMGE* Stockholm, Balkema PC, Vol 3 (1981), pp.677-682.
8. Gibson, R.E., Gobert, A. & Shiffman, R.L.: On Cryer's problem with large displacements and variable permeability. *Geotechnique* 40 (1990), pp.627-631.
9. Abousleiman, Y, Cheng, A. H.-D., Cui, L., Detournay, E. & Roegiers, J.-C.: Mandel's problem revisited. *Geotechnique* 46(2) (1996), pp.187-195.
10. Verruijt A.: The theory of consolidation. In: *Transport Phenomena in Porous Media* (Bear & Corapcioglu, ed.), Martinus Nijhoff, Dordrect, 1984, pp. 349-368.

Creep Tests on Reconstituted Soft Clay, Performed by Means of a Biaxial Device

ALESSANDRA BIZZARRI
Ministry of Transport, Public Works and Water Management, Delft, the Netherlands

Several creep tests on reconstituted soft clay were performed with a new biaxial apparatus designed at Technical University of Delft. With this device complex loading patterns can be applied, similar to field conditions. The aim of the tests is to study the relation between the creep behaviour and the applied stress path and stress configuration. The test results and the comparison with the creep behaviour described by different theoretical models (Janbu [1], Den Haan [2], Vermeer et al. [3], The et al. [4]) are presented in this paper.

Basic properties of the biaxial device

Only the basic features of the biaxial device are presented in this paper. A complete description was published before (Allersma [5], Bizzarri et al. [6]). The soil sample is confined by four rigid plates that can move in two directions: perpendicular and parallel to the sample surface. The first motion causes compression or expansion of the sample and the second is necessary to avoid collapsing between the plates. The displacement of each plate is realised by means of two actuators which are driven by step motors. Since the steps are defined accurately, the displacement can be deduced from the number of steps. The motion of the plates is synchronised electronically and by software. In the vertical direction the sample is confined by two fixed ridged plates. An image of the device is shown in Figure 1.

In order to eliminate all unknown friction components, the applied total stress is measured directly at the clay surface by means of four pressure sensors that are mounted in the plates. The pore pressure inside the sample is measured by means of a hypodermic needle combined with a pressure transducer. The probe is inserted in the centre of the sample from the bottom plate of the device. The control of the system and the data acquisition is performed by a computer. The biaxial apparatus can be driven in a stress or in a displacement control mode.

Test procedure

The tests were performed on reconstituted Dutch organic clay at stress levels that are comparable with the ones which occur in the field and similar to

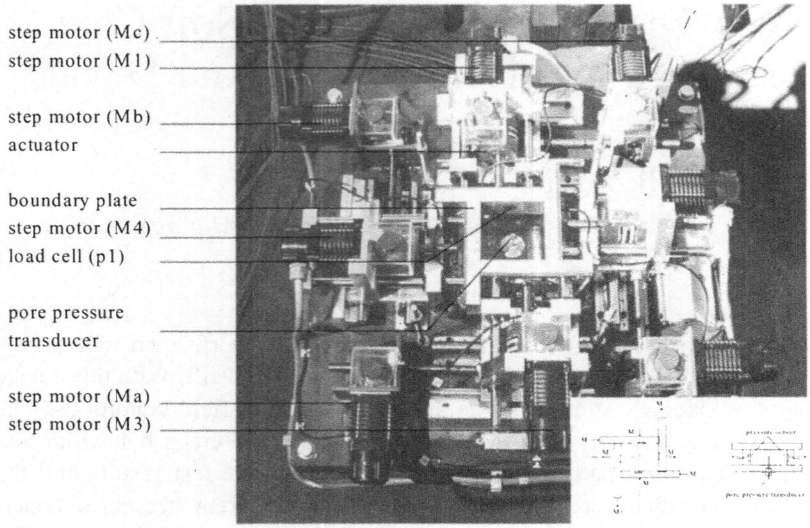

Figure 1. The biaxial device.

those resulting from simulations performed by Barneveld [7]. In order to study the creep behaviour of the clay, the samples were loaded to different stress configurations and consequently the total pressure applied to the clay surface was kept constant to investigate the strain-time relation. Two different loading procedures were used: a multiple stage loading procedure (MSL), as in a standard compression test, and a constant rate of strain (CRS) loading procedure.

Tests results

Typical results of a creep test are shown in Figure 2. Figure 2(a) shows the stress paths and final stress configurations of the performed creep tests. The time-dependent behaviour of the strain rate during secondary compression is shown in Figure 2(b). The volumetric and the deviatoric strain rate resulting are plotted as a function of $t'+\tau_0$. The dotted lines are fitted lines. The creep parameters μ_1*, μ_2*, μ*, t_r, and τ_0 are obtained from the experimental data according to the methods introduced by Den Haan [2] and Janbu [1] as depicted in Figure 2(c) and (d), respectively. The quantity t-t_r is the intrinsic time introduced by Janbu and Den Haan and is related to the loading history before the actual creep test. The quantity μ^*/τ_0 has the meaning of initial strain rate (Vermeer et al. [3]).

The time-dependent behaviour of the volumetric and the deviatoric strain rate resulting from the different creep tests (fitted lines) is shown in Figure 3(a) and 3(b), respectively. Although both quantities show a similar linear behaviour, the deviatoric strain rate is shifted towards larger values when

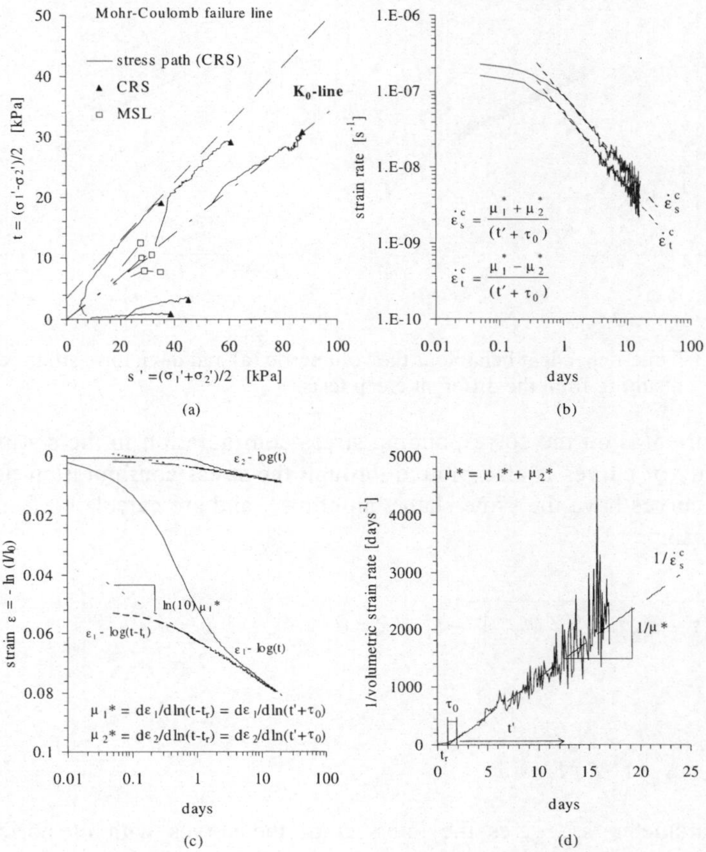

Figure 2. Typical creep test results. (a) Stress paths and tests configuration of the performed tests in the *s'*-*t* plane. (b) Volumetric and deviatoric strain rate. (c) Determination of secondary compression parameters (method Den Haan) (d) Determination of coefficient of secondary compression (Method Janbu).

corresponding to creep tests performed at a higher stress level. The stress level is here defined as the ratio between the applied deviatoric and average stress, *t/s'*. This behaviour of the deviatoric strain rate is in agreement with the observations made by Mitchell [8]. From results of drained triaxial tests on London clay and undrained tests on normally consolidated Sun Francisco Bay mud he concluded that the axial strain rate increases with increasing deviatoric stress. By plotting the volumetric and the deviatoric strain rates at a fixed time ($t+\tau_0$ = 1 day) as a function of the applied stress level (Fig. 4(a) and (b), respectively), it can be concluded that, while the deviatoric creep strain rate increases with the applied stress level, the volumetric creep strainrate is not dependent on the applied stress level.

The creep strain increment vectors resulting from the strain increments $\delta\varepsilon_s^c$ and $\delta\varepsilon_t^c$ and obtained from the different creep tests are superimposed

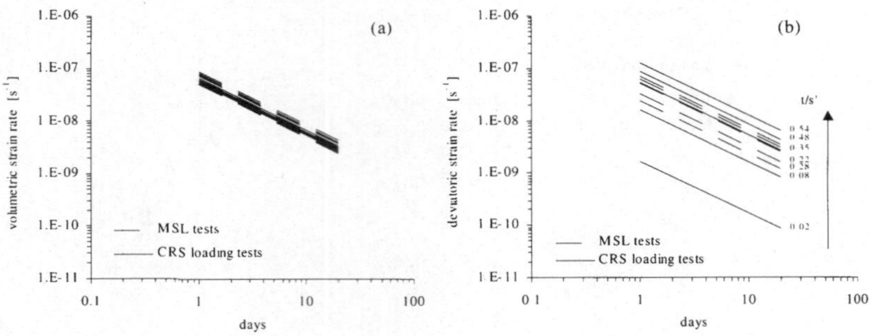

Figure 3. Time-dependent behaviour the volumetric (a) and deviatoric strain rate (b) resulting from the different creep tests.

in Figure 5(a) on the corresponding stress configuration in the s'-t plane. A family of curves is also plotted through the stress configuration points. These curves have the same shape, an ellipse, and are calculated from the relationship

$$\sqrt{(s' - s_{eq})^2 + (t / M_{CS})^2} - s_{eq} / 2 = 0 \tag{1}$$

with

$$s_{eq} = s' + t^2 / (M_{CS}^2 s') \tag{2}$$

The parameter s_{eq} gives the intersect of the curves with the horizontal axis t = 0. The parameter M_{cs} indicates the slope of the critical state line and defines the shape of curves. The direction of the creep strain increment vectors is roughly normal to the family of curves defined by M_{cs}= 0.979. In Figure 5(b) the deviation from normality is plotted against the stress level. A maximum deviation of ± 10 ° was found. The points denoted with the symbol ' ∗ ' relate to plastic strain increments corresponding to stress configurations during the CRS loading.

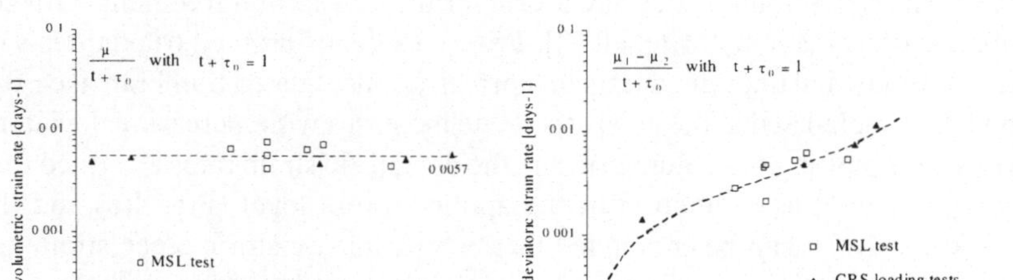

Figure 4. Volumetric (a) and deviatoric (b) strain rate at a fixed time.

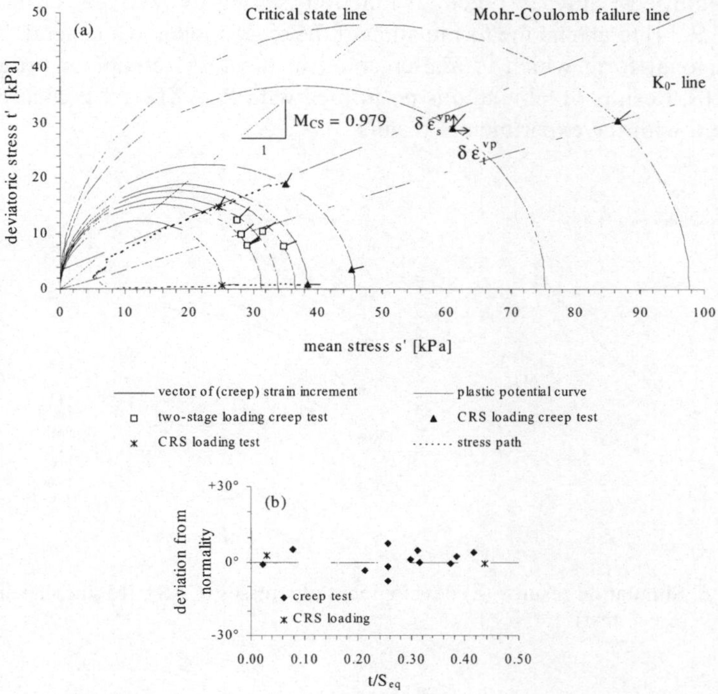

Figure 5. (a) (Creep) strain increment vectors at the different stress configurations and family of potential curves. (b) Deviation from normality to the potential curves.

Equation (1) defines the Modified Cam Clay plastic potential function (Roscoe & Burland [9]) in the plain strain representation. It can be concluded that the creep behaviour in the plane strain approximation (s'-t plane) conforms to an associated flow rule similar to the one described by the Modified CamClay model. The potential curve can be described by an ellipsis. The slope of a vector normal to the potential curve is given in each point by

$$\Delta t / \Delta s = (M_{CS}/2)\left[(s_{eq} - 2s')/\sqrt{s's_{eq} - s'^2}\right] \tag{3}$$

The deviatoric strain rate vector can be then written as:

$$\dot{\varepsilon}_t^c = (M_{CS}/2)[(s_{eq} - 2s')/\sqrt{s's_{eq} - s'^2}\,]\dot{\varepsilon}_s^c \tag{4}$$

Figure 4(b) shows a good agreement between the experimental results and the curve obtained from equation 4 with $\dot{\varepsilon}_s^c = 0.0057$.

The Modified CamClay potential function is used by Vermeer et al. [3], The et al. [4] to extend the formulation of the creep model to a general three-dimensional form which is implemented in the FEM computer program PLAXIS. Results of simulations performed with PLAXIS are compared in Figure 6 with the experimental results.

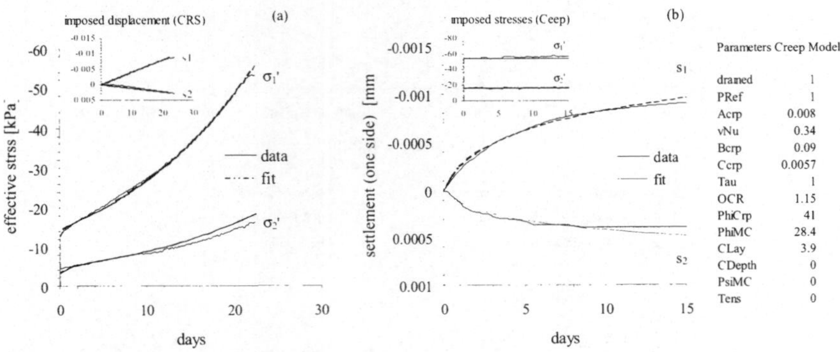

Figure 6. Simulation results: (a) development of stresses (CRS), (b) displacements (creep test).

References

1. Janbu, N: The resistance concept applied to soils. *Proc. of the 9th ICSMFE*, Tokyo, Vol 1 (1969), pp.289-292.
2. Den Haan, E.J: Generalization of the stress, strain and time measures of soil compression. *Proc. Int. Symp. on Compression and Consolidation of Clayely Soils.* Hiroshima, Vol 1 (1995), pp.513-518.
3. Vermeer, P.A., Stolle, D.F.E. & Bonnier, P.G.: From the classical theory of secondary compression to modern creep analysis. *Proc. 9th Int.Conf. on Computer Methods and Advances in Geomechanics,* Wuhan China, Vol 4 (1997).
4. The, B.H.P.A.M., Vermeer, P.A. & Termaat, R.J.: A viscoplastic creep model for the engineering practice. *Proc. Int. Symp. On Problematic Soils*. Sendai, Vol 1 (1998), pp.657-660
5. Allersma, H.G.B.: *New design of a biaxial apparatus for soft soils*. Delft University of Technology 1993. Report no. 354.
6. Bizzarri, A., Allersma, H.G.B. & Koehorst, B.A.N.: Preliminary tests on soft clay with a biaxial apparatus. *Proc. Int. Symp. on Compression and Consolidation of Clayey Soils*, Hiroshima, Vol 1 (1995), pp.27-32.
7. Barneveld, A.S.:*Spanningspaden behorende bij verschillende dijk(ophoog) profielen* (deel2). Report RWS-Dienst Weg en Waterbouwkunde 1995.
8. Mitchell, J.K. *Fundamentals of soil behaviour.* 2nd ed., John Wiley & Sons, Inc., New York 1993.
9. Roscoe, K.H. & Burland, J.B.: On the generalised stress-strain behaviour of 'wet' clay. *Engineering Plasticity* (eds. Heyman, Leckie), Cambridge University Press (1968), pp.535-609.

Salt Deformation and Subsidence

Peter A. Fokker, Gerard A.M. Kruse
NEDMAG Industries, Veendam / BECi, Vries
GeoDelft, Delft, the Netherlands

Subsidence is an issue that requires ever more attention, both from a technical and a socio-political point of view. Subsidence is not only the result of compaction of soil or rock materials (as is the case for dewatering clay or peat and depressurzing gas-bearing porous rocks), but also of the active mining of solid material from the underground. Active removal of material results in an elastic response of the surrounding rocks and soils, but usually also in a visco-plastic response of the loaded environment, either in a non-destructive or destructive way.

Solution mining in deeper salt layers (>1 km depth) results in a measurable salt flow towards the cavern (brine filled volume which is the result of dissolving the salt). This salt flow is accompanied by surface subsidence.

Salt is a rock material that deforms easily by so-called creep when loaded by deviatoric stresses. The deformation occurs mainly at constant volume and is usually non-destructive, comparable to the hot deformation of metals or wax. Like with wax, the deformation rate of salt is temperature dependent. Salt stresses in undisturbed deep salt are always (almost) isotropic as a result of the tendency to deform easily.

To predict subsidence and related changes above salt caverns one needs to know

- the amount of salt flow towards a cavern (cavern convergence) as a function of size, depth, and cavern pressure;
- the relation between convergence and the depth and shape of its subsidence bowl;
- the effects of multiple subsidence sources (multiple caverns, combined salt and natural gas extraction, etc).

Both numerical and semi-analytical models are required to predict subsidence and related changes with sufficient accuracy.

Two examples of Dutch salt extraction are discussed: the sodium salt extraction of FRISIA near Harlingen and the magnesium salt extraction of NEDMAG near Veendam.

Data acquisition

In theory it should be possible to predict subsidence with Finite Element computations. Suitably accurate and reliable data on stiffness and creep parameters of all rock-soil layers involved, and sufficient computing power (to perform 3-D calculations of a 3-D cavern field) are usually not available. Actual field data on subsidence and salt extraction are available, however, enabling extraction of data from field tests, by tuning results of FE-element computations to the observations. Superposition models can than be applied to predict the long-term behaviour of the cavern field and the effects on the surface.

Important measurements or data recordings of the field tests are:

- levelling excersizes;
- production data (produced salt, injected water, pressures, etc);
- cavern size measurements (SONAR-measurements).

Levelling exersises give the input on the shape of the subsidence bowl. Effects of interfering subsidence from convergence of multiple caverns, or other sources of subsidence, have to be extracted from the levelling excersise before the relation between subsidence and convergence of a single cavern can be found. The shape of a single-source subsidence bowl, defined as a cross section through the bowl's centre, appears to be quite similar to a Gaussian (or Bell) type of curve (Fig. 1). Both the results of FE computations and the levelling exersizes can be fitted well with a Gaussian type of equation for subsidence w.

$$w(x,y) = w_{\max} \exp(-\gamma r^{\delta}) = \chi V_{con} \exp(-\gamma r^{\delta})$$

The parameters γ and δ describe the curvature of the subsidence bowl as a function of distance r from the cavern axis, and V_{con} is the volume of the cavern convergence. The maximum subsidence can be related to the convergence volume of the cavern via parameter χ. The superimposed subsidence field is (for caverns 1 to n)

$$w(x,y) = \sum_{i=1}^{n} \chi_i V_{con,i} \exp(-\gamma_i r_i^{\delta_i})$$

Table 1. Subsidence parameters NEDMAG and FRISIA.

parameter	NEDMAG	FRISIA
γ (-)	$6.0 * 10^{-6}$	$4.5 * 10^{-7}$
δ (-)	1.7	1.97
χ (m^{-2})	$0.96 * 10^{-7}$	$1.0 * 10^{-7}$

It appears that the parameters are virtually indepent of time during operation, which is also supported by Finite Element studies [8]. There is a slight trend (after closure of the cavern field) to a flattening of the bowl during the first decades, but the effect can be neglected in the cases considered here.

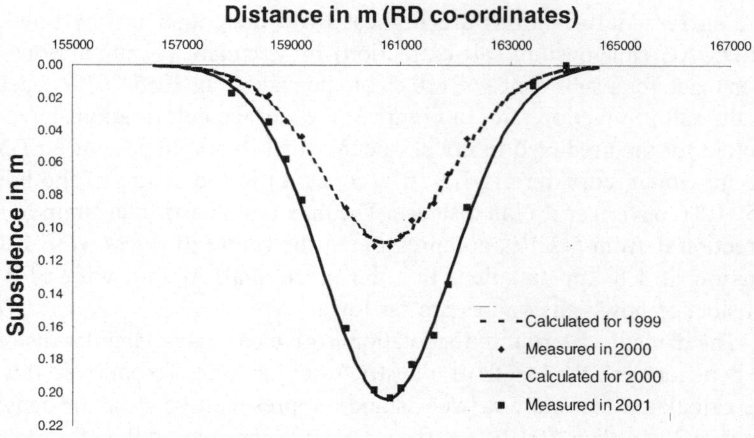

Figure 1. Levelling results; salt extraction at FRISIA salt, Harlingen; a Gaussian fit.

FE-calculations

By modifying the elastic and creep input-data in the FE-model within the range of uncertainty, one can fit the observed cavern and subsidence behaviour (Fig.2). The fitted FE-model can then be used to extrapolate subsidence data and to calculate stresses and deformations like horizontal strain and bending at the surface for higher salt extraction.

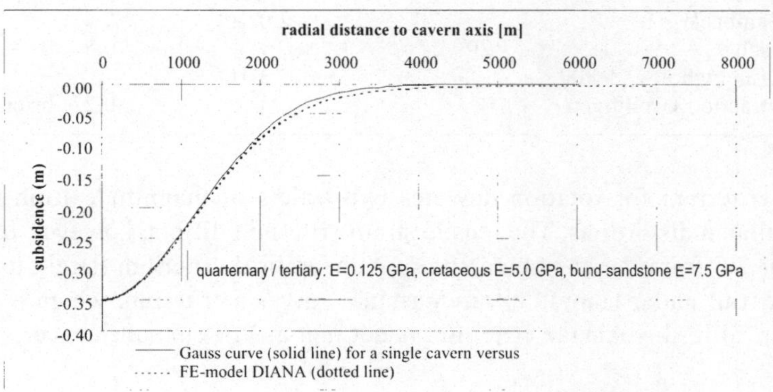

Figure 2. FE-subsidence for salt extraction at FRISIA salt, with the corresponding by a Gaussian curve.

Damage to constructions on the surface is notably related to angular distortion (bending), horizontal strain, and tilt. The results from the FE-modelling can be compared to damage criteria for constructions, as postulated in the literature.

The surface deformations are highest for a deep and narrow bowl. For NEDMAG (magnesium salt extraction) in Veendam [7] predictions have been made for a subsidence of 110 cm, to be realised in 2035-2050, depending on the salt production rate. In Figure 3, the surface deformations have been plotted for the predicted future subsidence of 110 cm. In case of NEDMAG the maximum curvature is 1.3 10^{-6} (concave) in the centre of the bowl to 0.5 10^{-6} (convex) at 1.5 km distance. The maximum horizontal strain in radial direction is from 5.2 10^{-4} compression at the centre of the bowl to 1.8 10^{-4} tension at 1.8 km distance. For the more shallow and wider FRISIA subsidence bowls the values are far lower.

The damage criterium β for rotation is given as a total angular distortion, i.e dependent on the length of the structure: $\beta = \frac{1}{2}\kappa L$. To compare this with the calculated curvature κ, we assume a representative structure length. If we take 20 m, then $\beta = 10\kappa$, i.e. from 1.3 10^{-5} (concave) to 0.5 10^{-5} (convex).

Damage criteria

Damage criteria for buildings in the literature are given in Table 2.

Table 2. Damage criteria for subsidence related effects (see Ref. list)

	radius of curvature [km]	relative rotation β	strain [ε] extension in promilles
Bjerrum		1.2 10^{-3}	
Boscardin		1.0 10^{-3}	0.5
van Sambeek		2.0 10^{-3}	0.5
Kratzsch	20		0.5
NEN (nieuwbouw) 1:300		3.3 10^{-3}	
gasextraction Groningen			0.25; brick.

The criterium for rotation deviates two orders of magnitude from from calcultated distortions. The tensile strain criterium differs, however, only a factor 2.5 from the computed strain. A theoretical criterium (strain test on brick wall under laboratory circumstance) gives a criterium which is 50% lower. Still, the damage criterium is not met at 110 cm subsidence.

Other damaging effects

Aside from the direct influence of the caverns on the surface, the indirect effect of subsidence on the water table may be more important for the integrity of structures. If the water table is not adjusted to follow the subsidence, or is overadjusted, a relatively higher or lower water table results. Especially constructions on shallow fundations and weak top soils (containing clay and peat) may experience damage by compaction of these layers if the water

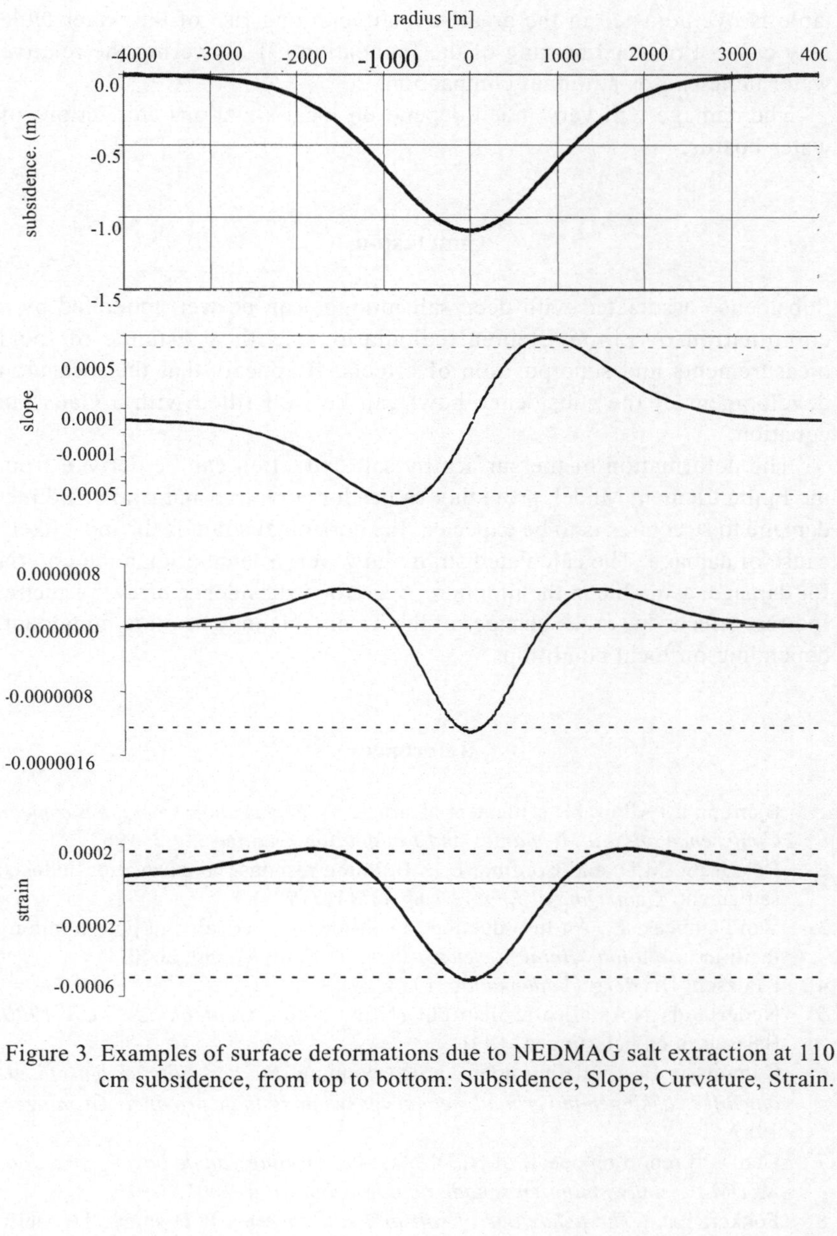

Figure 3. Examples of surface deformations due to NEDMAG salt extraction at 110 cm subsidence, from top to bottom: Subsidence, Slope, Curvature, Strain.

table is overadjusted in the area. A relative strong rise of the water table may cause a reduced bearing of the foundation [9]. Lowering the relative water table causes aditional compaction.

The damage will very much depend on local situations and actions of water boards.

Conclusions

Subsidence associated with deep salt mining can be well modelled by a combination of Finite Element calculations, with assistance of local measurements and superposition of effects. It appears that the measured development of the subsidence bowl can be well fitted with a Gaussian equation.

The deformation of the surface by salt extraction can be derived from the Finite Element Model, providing values for curvature and strain. If direct damage to structures is to be expected, the horizontal strain is the most likely cause of damage. The calculated strain, however, is smaller than the criteria for damage, available in the literature, even for a subsidence of over 1 metre. Indirect effects due to the changes of the water table may be more important, depending on local conditions.

References

1. Bjerrum, L.: Allowable settlement of structures. *Proceedings from the European Conference on Soil Mechanics and Foundation Engineering 2*, 1963.
2. Boscardin, M.D. and Cording, E.J.: Building response to excavation-induced settlement. *Geotechnical Engineering* 115(1) (1989).
3. Van Sambeek, L.: An Introduction to subsidence over salt and potash mining facilities. *Solution Mining Research Institute Fall meeting*, 2000.
4. Kratzsch, H.: *Bergschadenkunde*, 1974.
5. Nederlands Normalisatie-instituut (NEN 6740): *Geotechniek TGB 1990*, Basiseisen en belastingen, 1991.
6. Commissie Bodemdaling door Aardgaswinning: *Studieresultaten betreffende ongelijke zakkingen in verband met aardgaswinning in de provincie Groningen*, 1987.
7. GeoDelft report on behalf of NEDMAG: *Vervorming van de bovengrond door NEDMAG zoutwinning en schade aan de bebouwing*, 2001,
8. Fokker, P.A.: *The behaviour of salt and salt caverns*, PhD study, TU Delft, 1995
9. Verruijt, A.: Schade bij gelijkmatige bodemdaling, *Geotechniek* 15 (2000)

The Influence of Creep Ageing on $K_{o,nc}$

EVERT J. DEN HAAN
GeoDelft, Delft, the Netherlands

A 2-D generalization of the *a,b,c* - isotache model [1] is used to study how the normally consolidated value of K_o is affected by creep. Expressions are derived for $K_{o,nc}$ during constant rate of strain loading, and during creep at constant vertical effective stress. Creep from a normally consolidated state can result in both increasing and decreasing $K_{o,nc}$ depending on the relative values of a number of soil parameters. This issue was intensely debated and researched in the 1980's ("Does $K_{o,nc}$ increase or decrease during secondary compression?"). An increase or constancy was most opted for, some work indicated a decrease, but the view that all options could occur was absent.

The model

The model used here is essentially that published by Kutter and Sathialingam [2]. The finite element package Plaxis [3] uses a similar model for soft, creeping soil, but it is was conceived as a 2-D generalization of the *a,b,c* - isotache model [1]. The model is illustrated below. An axially symmetric formulation suffices in the present case.

Logarithmic or natural strain is used. Stresses are effective throughout. Principal stress and strain directions are vertical and horizontal. The subscripts $.._a$ and $.._r$ denote the vertical (axial) and horizontal (radial) directions. The subscripts $.._p$ and $.._q$ denote isotropic and deviatoric strain quantities. Deviatoric stress $q=\sigma_a-\sigma_r$ and deviatoric strain $\varepsilon_q=2(\varepsilon_a-\varepsilon_r)/3$. The superscripts $..^d$ and $..^s$ denote direct and secular strains and strain rates. These latter terms are synonymous with the terms "elastic" resp. "creep" or "visco-plastic" used in other models and are due to Keverling Buisman who was among the very first to observe and study creep in soils.

Expressions are derived for direct and secular strain rates as functions of the present state. Total strain rate is the sum of both, and strain can be obtained by integration in time.

The relationships for elasticity make use of an isotropic stiffness which is linearly dependent on the isotropic stress p, and a constant Poisson's ratio. That is, isotropic stiffness equals p/α and shear modulus equals $p/(2\alpha\beta)$ where $\beta=(1+\nu)/(3(1-2\nu))$. Then

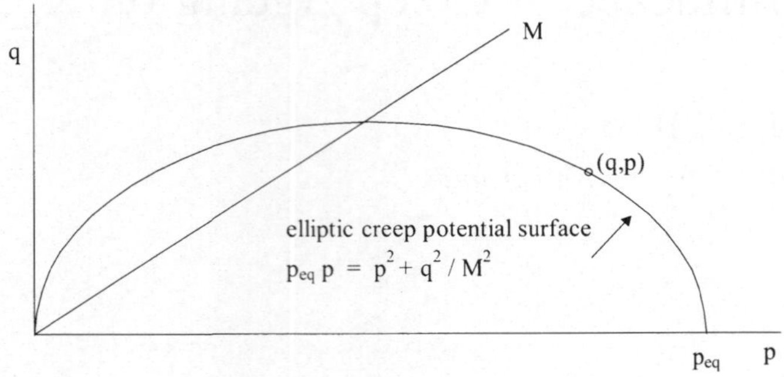

Figure 1. Creep potential surface

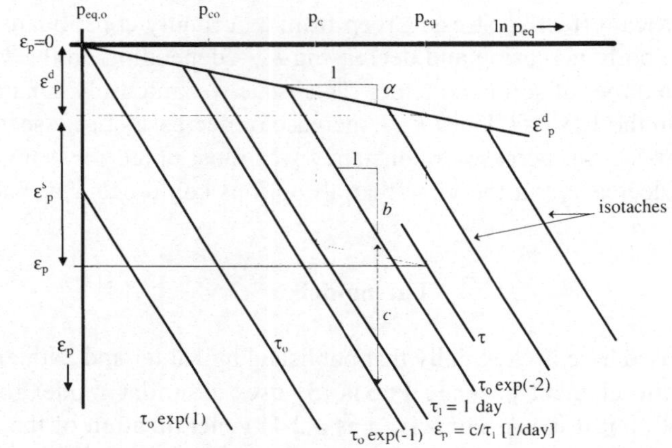

Figure 2. Isotaches in the isotropic stress - strain plane

$$\begin{aligned} \dot{\varepsilon}^d_p &= \alpha \;\; (\dot{\ln p}) = \alpha \;\; \dot{p} / p \\ \dot{\varepsilon}^d_q &= 2\alpha \beta \dot{q} / 3p \end{aligned} \qquad (1)$$

The visco-plastic rates of strain are derived from the expression

$$\dot{\varepsilon}^s_{ij} = \frac{c}{\tau_1}\left(\frac{p_{eq}}{p_c}\right)^{\frac{b-\alpha}{c}} \frac{\partial p_{eq}}{\partial \sigma_{ij}}$$

with (2)

$$p_c = p_{co}\,\exp(\varepsilon^s_p / (b-\alpha))$$

For isotropic conditions with $q=0$, equation (2) reduces to

$$\dot{\varepsilon}_p^s = \frac{c}{\tau_1}\left(\frac{p_{eq}}{p_c}\right)^{\frac{b-\alpha}{c}} \tag{3}$$

and this equation is visualized in Figure 2, together with (1a). Total strains are used in the figure to maintain the likeness to measured data. The component direct and secular strains are also shown. p_{eq} is an equivalent stress defined by the stress state and the critical state parameter M as in Figure 1:

$$p^2 + q^2/M^2 = p_{eq}\,p. \tag{4}$$

This is an ellipse of the same form as that of modified Camclay, but it is now a visco-plastic potential surface rather than a yield surface. c/τ_l is a reference rate of compression where τ_l equals 1 day. p_c is the stress on the reference isotache at the current visco-plastic isotropic strain. An isotropic stress-strain state yields a value of τ by interpolation from the isotaches, and rate of secular isotropic strain follows from c/τ. Equation (3) results.

From the above equations, the strain rates in the vertical and horizontal directions follow:

$$\begin{aligned} \dot{\varepsilon}_a^d &= \frac{\alpha \dot{p}}{3p}\left(1+2\beta\frac{\dot{q}}{\dot{p}}\right) & \dot{\varepsilon}_a^s &= \frac{c}{3\tau_1}\left(\frac{p_{eq}}{p_c}\right)^{\frac{b-\alpha}{c}}\left(1-\frac{\eta^2-6\eta}{M^2}\right) \\ \dot{\varepsilon}_r^d &= \frac{\alpha \dot{p}}{3p}\left(1-\beta\frac{\dot{q}}{\dot{p}}\right) & \dot{\varepsilon}_r^s &= \frac{c}{3\tau_1}\left(\frac{p_{eq}}{p_c}\right)^{\frac{b-\alpha}{c}}\left(1-\frac{\eta^2+3\eta}{M^2}\right) \end{aligned} \tag{5}$$

where $\eta = q/p$ is the stress ratio.

K_0 equations

During one dimensional compression at constant rate of strain C the strain rate components of equation (5) are related by

$$\begin{aligned} \dot{\varepsilon}_a &= \dot{\varepsilon}_a^d + \dot{\varepsilon}_a^s = C \\ \dot{\varepsilon}_r &= \dot{\varepsilon}_r^d + \dot{\varepsilon}_r^s = 0 \end{aligned} \tag{6}$$

During steady state deformation, in the normally consolidated range, each of these components is constant. Then, the factors in the expressions for the direct strain rates, $\dot{p}/p$ and $\dot{q}/\dot{p}$, are also constant, resulting in $\eta_{Ko} = \dot{q}/\dot{p} = q/p$.

Using equation (6), the power factors in (5) can be eliminated, yielding

$$\frac{\dot{p}}{p} = C\left[\frac{\alpha}{3}\left(1+2\beta\,\eta_{Ko,nc}\right)\frac{\alpha}{3}\left(1-\beta\,\eta_{Ko,nc}\right)\frac{M^2-\eta_{Ko,nc}^2+6\eta_{Ko,nc}}{M^2-\eta_{Ko,nc}^2-3\eta_{Ko,nc}}\right]^{-1} \tag{7}$$

The stress - strain state in Figure 2 will follow an isotache with slope b because of the constant rates. It follows that

$$\dot{\varepsilon}_a = b\,\dot{p}\,/\,p \tag{8}$$

and consequently

$$b = \frac{\alpha}{3}\left(1+2\beta\,\eta_{Ko,nc}\right)-\frac{\alpha}{3}\left(1-\beta\,\eta_{Ko,nc}\right)\frac{M^2-\eta_{Ko,nc}^2+6\,\eta_{Ko,nc}}{M^2-\eta_{Ko,nc}^2-3\,\eta_{Ko,nc}} \tag{9}$$

also to be written as

$$\eta_{Ko,nc}^3 - \frac{b}{\alpha\,\beta}\eta_{Ko,nc}^2 - (M^2+3\frac{b}{\alpha\,\beta}-\frac{3}{\beta})\eta_{Ko,nc} + M^2\frac{b}{\alpha\,\beta} = 0 \tag{10}$$

$K_{o,nc}$ follows from

$$K_o = (3 - \eta)/(3 + 2\eta) \tag{11}$$

The applied rate of strain *C* does not appear in equation (10), making (10) valid for any sequence of different values of *C*. An inviscid version of the model, in which the visco-plastic potential surface is used as a yield surface, would yield the same expression, see [3]. This makes the equation rather general. If *C=0* however, relaxation occurs and *q* and *p* gradually decrease. This case is not considered here.

At constant vertical stress, ageing occurs and the lateral stress reaches an ultimate value. Consequently, direct, elastic deformation ceases. Only the secular deformation persists. Therefore $\alpha = 0$ in equation (10), and the ultimate state after ageing is given by

$$\eta_{Ko,\infty}^2 + 3\eta_{Ko,\infty} - M^2 = 0 \tag{12}$$

from which

$$K_{o,\infty} = \frac{9}{2\sqrt{9+4M^2}} - \frac{1}{2} \tag{13}$$

The rate of change of K_o during ageing follows from the conditions that rate of change of the axial stress is zero, and radial strain is zero. Combined with equation (5) it emerges that

$$\dot{\varepsilon}_a = \frac{3\alpha\,\dot{p}}{2p}\,\frac{-\beta\eta^2 + 2\eta + \beta M^2}{\eta^2 + 3\eta - M^2} \qquad (14)$$

For numerical solution purposes, the time increment associated with an increment of strain can be found from

$$\Delta t = \frac{\Delta\varepsilon_p - \Delta\varepsilon_p^d}{\dot{\varepsilon}_p^s} = \frac{\Delta\varepsilon_a - \Delta\varepsilon_p^d}{\dot{\varepsilon}_p^s} = \frac{\dfrac{3\alpha\,\Delta p}{2p}\,\dfrac{-\beta\eta^2 + 2\eta + \beta M^2}{\eta^2 + 3\eta - M^2} - \dfrac{\alpha\,\Delta p}{p}}{\dfrac{c}{\tau_1}\left(\dfrac{p_{eq}}{p_c}\right)^{\frac{b-\alpha}{c}}\left(1 - \dfrac{\eta^2}{M^2}\right)} \qquad (15)$$

which allows Δt to be determined from Δp or vice versa.

Evaluation

Equation (12) corresponds to modified Camclay as presented in [4] in the absence of elasticity. In [2], similar equations are given as an approximate solution for one-dimensional compression, by assuming elastic strains to be much smaller than the visco-plastic strains. We see now however that distinction should be made between the normally consolidated value of K_o as it can be found from equation (10) and the ultimate, aged value from equation (12).

Figure 3 depicts (10) and (12), showing $K_{o,nc}$ and $K_{o,\infty}$ as function of M. Jáky's correlation is shown as reference. (10) is a complex equation. For a given value of b/α there is a range of curves for different ν values. The width of the range decreases for increasing b/α. For practical values of M and ν, say $M<1.5$ and $0.05<\nu<0.35$, $K_{o,nc}$ increases during ageing from the normally consolidated condition, and this agrees with the majority of the work done on this issue in the 1980's. However, higher values of ν and M can produce an opposite tendency. For example, this would occur with ν=0.35 for M>1.6.

Dutch organic clays have been found to have high values of M and low $K_{o,nc}$ (in addition to strong compressibility and creep sensitivity). E.g., Oostvaardersplassen clay with an organic content of ~20%, has M values of above 2 and $K_{o,nc}$ values (measured in a K_o oedometer at constant rate of strain) of 0.4 or less. This may plot above the "ultimate" curve in Figure 3, and, considering that the model used here generally fits measured behaviour of this clay well, $K_{o,nc}$ could well decrease during creep ageing.

Admittedly, the model need not correctly reflect soil behaviour, and the (combination of) parameter values which plot above the ultimate line is

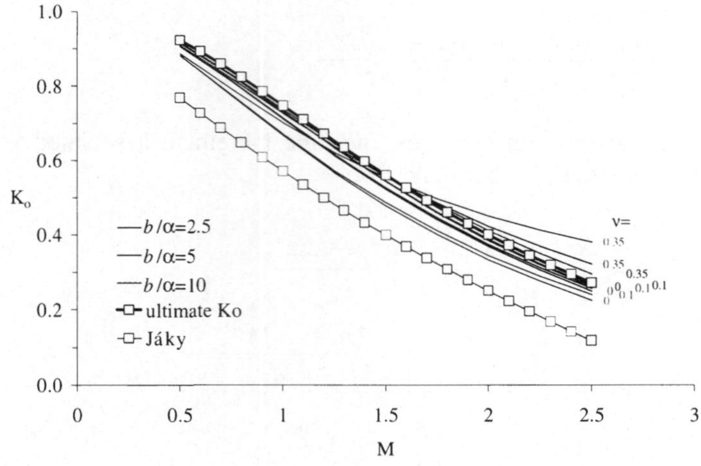

Figure 3. Ko,nc as function of M, ν and b/α

possibly rare or even unlikely. Certainly the arguments in the literature (see e.g. the discussions of [5]) for increasing $K_{o,nc}$ are persuasive. But some test results do point in the direction of a decrease, and this paper shows both could be possible.

The "behavioural reasoning", [5], that the soil state will tend to K_o=1 over geologic time due to gradual failure of particle bonds if a deviator stress is active, is not predicted by the model. Nor do laboratory and field tests indicate this to occur.

The rate with which $K_{o,nc}$ approaches the ultimate value can be derived from equations (14) and (15). For the Oostvaardersplassen clay, the model revealed very quick changes, and the ultimate value is reached within 1 day.

The ultimate value of equations (12) and (13) also applies to overconsolidated states. Generally K_o decreases during ageing from overconsolidated states, and the associated time scale is much larger than for normally consolidated states. The model effectively predicts constant K_o during ageing from substantially overconsolidated states.

References

1. Den Haan, E.J.: A compression model for non-brittle soft clays and peat. *Géotechnique* 1(1996), pp.1-16.
2. Kutter, B.L. & Sathialingam, N.: Elastic-viscoplastic modelling of the rate-dependent behaviour of clays. *Géotechnique* 3(1992), pp.427-442.
3. *Plaxis Material Models Manual*, Version 7. Ed. R.B.J. Brinkgreve & P.A. Vermeer. A.A. Balkema, Rotterdam, 1998.
4. Roscoe, K.H. & Burland, J.B.: On the generalized stress-strain behaviour of 'wet' clay. Conference papers, *Engineering Plasticity*. Ed. J. Heyman & F.A. Leckie. Cambridge at the University Press, (1968), pp.535-609.
5. Kavazanjian, E. & Mitchell, J.K.: Time dependence of lateral earth pressure. *J. Geot. Eng.* ASCE. 4(1984), pp.530-533.

Uncertainties in Settlement Predictions

JACQUES P. OOSTVEEN
TUDelft, Delft, the Netherlands

By means of interviews the steering committee on *Soil and Foundation* of the *CROW* made an inventory of future research needed for the road-building sector [1]. One of the problems highlighted by this inventory is the uncertainty in the prediction of settlements. There is a need to get more insight into the natural variation of the subsoil. Especially, about how sensitive the predictions are to the natural variation, as well as about the other causes of the mispredictions. To look into this matter a working committee has been installed. This paper presents some results.

The inventory shows a general and fundamental problem between theory and practice. The predictability of settlements is related to our knowledge of soil behaviour and to the variation of the subsoil. Therefore, first a short overview is given about the prediction methods and secondly some aspects of the soil behaviour and its variation are discussed.

Prediction methods

Predictions of settlements are based on curve fitting using the theoretical interpretation of the compression behaviour of soil. Terzaghi [2] formulates the logarithm law of compression, including the time dependent behaviour by the process of consolidation. Keverling Buisman [3] added secondary settlements, linear on a logarithm time scale to the primary settlements. After Koppejan [4] combined these two theories using the principles of superposition, the resulting theory became most common to determine settlements in the Netherlands. In formula, it reads

$$\varepsilon = \left[\frac{1}{C_p} + \frac{1}{C_s} .\log\ t \right] .\ln\left(\frac{p_2'}{p_1'}\right) \qquad (1)$$

Taylor [5] defined an expression of the secondary settlements, without the principle of superposition. This theory is used worldwide.

$$\varepsilon = \frac{C_c}{1+e_0} \ln\left(\frac{p_2'}{p_1'}\right) + C_a \log(t) \qquad (2)$$

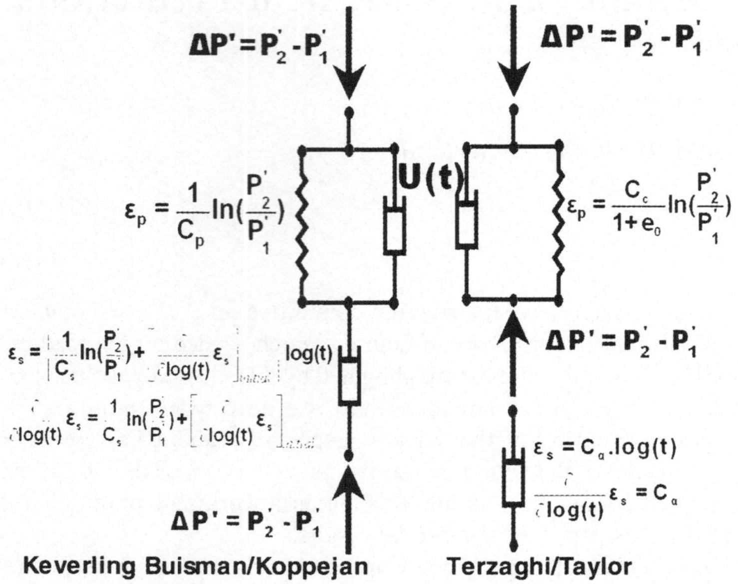

Figure 1. Reologic models of both schools.

These two theories represent different schools for settlement prediction. Terzaghi/Taylor considered each loading step as a new test, while Keverling Buisman/Koppejan considered each loading step as added to the prior step by superposition. Terzaghi/Taylor added the load independent secondary settlements at the end of the consolidation period, while Keverling Buisman/Koppejan added load dependent secondary settlements to the primary settlements by superposition, immediately after the first effective stresses were developed.

Figure 1 shows the reologic models of both theories. Because of the interpretation for which each loading step is considered as a new test the method of Terzaghi/Taylor disregards the initial rate of settlement due to the prior step. If a very small loading step is applied, the initial rate of settlement of the prior loading step mainly determines the total rate of settlement. By disregarding this aspect, it will be impossible to formulate a load dependent expression of the secondary settlement. From the reologic point of view it seems not logic.

From these theories including recent modifications the constitutive parameters can determine by curve fitting. Therefore, two straight lines schematize the relation between compression and the load at logarithmic scale, which is curved. The stress belonging to the crossing point of these two branches is p_c. The first branch outlines the relative stiff soil behaviour below p_c, while the second line beyond p_c represents the weaker behaviour of virgin soil.

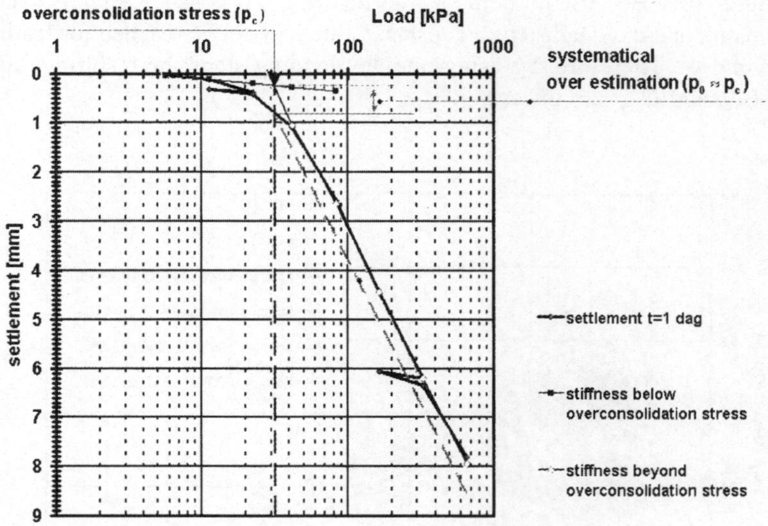

Figure 2. Systematically error.

Normally, for curve fitting, boundaries are very important and they have to be chosen such that the interpretation can be done by interpolation and not by extrapolation. Therefore, the time and load boundaries have to be determined accurately by design loads. However, Figure 2 shows another kind of systematical error. At the level of p_c curve fitting will show large deviation. In case of rather virgin soil this error can be relatively large. To reduce this systematical error the load boundaries, p_0 and $p_0+\Delta p$, have to be fitted well. It is suggested to fit this part of the curve by taking a parabolic expression or by several branches in accordance with the loading steps. It should not be a problem.

From Figure 2 it will be clear that the fitted value below pc may vary significantly. Depending on the chosen magnitudes of the first two loading steps the first branch will be positioned on different locations of the curve. Because beyond p_c the logarithmic load-settlement relation is rather linear, this branch will fit better. Figure 3 shows some results from a soil investigation for a 20 km road construction. The wet weight has been used as characteristic parameter for the soil type.

From the variation of the primary compression coefficients determined by the method of Keverling Buisman/Koppejan, it appears that beyond p_c $1/C'_p$, which is rather equal to $C'_c/(1+e_0)$, is normally distributed with a standard deviation of about 0.024. In accordance to the method of curve fitting the variation of $1/C_p$ (below p_c) has not a constant standard deviation.

The same can be said for $1/C_s$ and $1/C'_s$. However, for both theories the time boundary is very important to determine the secondary compression coefficients. While in former days the oedometer test was performed by a 7 days loading step, at present a one-day loading step is generally used, to reduce costs. Because for both theories *dS/d(log t)* is constant, Figure 4 shows by using one-day load steps a constant value will not be reached for loadings beyond p_c. Therefore, by using one-day loading steps the resulting values of C'_s and C'_α are uncertain.

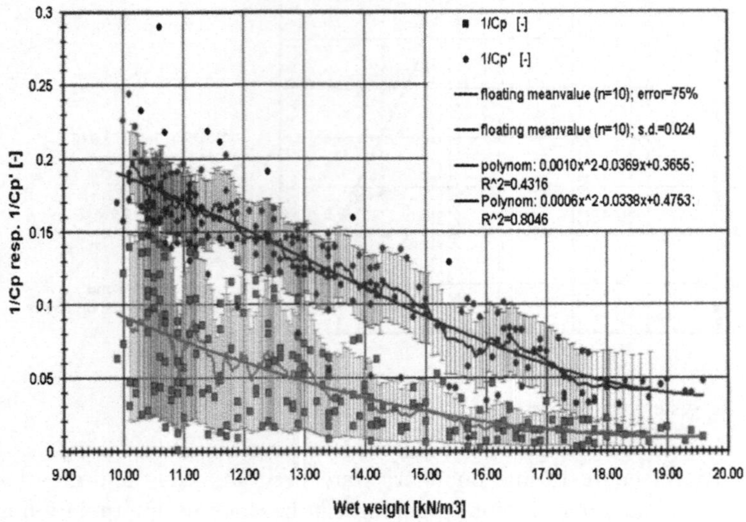

Figure 3. Variation of C_p and C'_p versus wet weight.

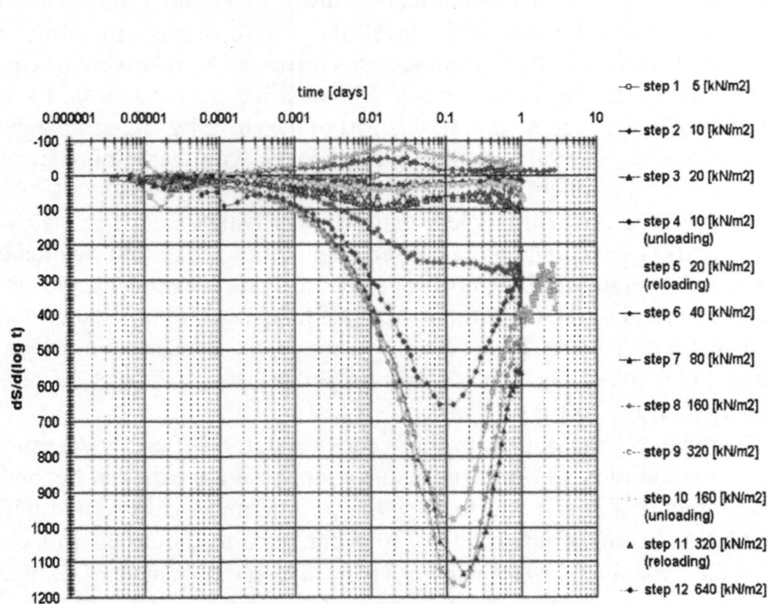

Figure 4. Velocity of settlement.

Evaluation of the soil behaviour under compression

A better prediction of the settlements needs a better insight into the behaviour of the soil during compression. Although there is a lot of experiences by the traditional methods, inherent to these methods there are systematical errors. Errors by taking too little time for the loading steps, by not taking into account that the compression coefficient and p_c are fitting parameters, correlated and determined by the applied loading steps, as well as by a wide variation of c_v and the determination of p_0.

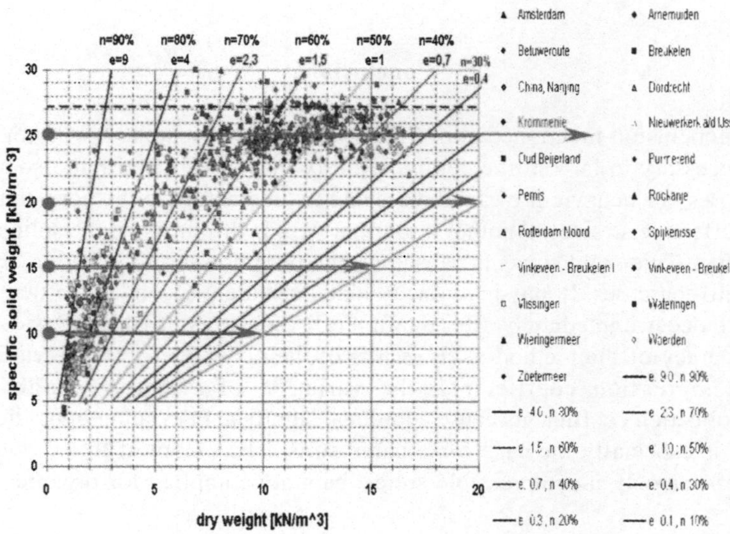

Figure 5. Dry weight versus specific solid weight.

In Figure 3 the wet weight has been used to characterize the soil type. In fact, it is one of the three primary charactistic parameters from which the other identities like water content, porosity, void ratio as well as the rate of saturation can be calculated. Figure 5 shows the relation between two other primary characteristic parameters, the dry weight of the soil and the specific solid weight, by comparing measured data to the theoretical relation between these parameters, as well as the porosity and the void ratio. These data show a curved path from a dry weight of 1 kN/m^3 at a specific solid weight of about 5 kN/m^3 along the relation for about n=80% to a dry weight of about 17 kN/m^3 at a specific solid weight of about 27 kN/m^3. It is clear that dependent upon the amount of organic material the specific weight of the solids will differ from the specific weight of the mineral material of about 27 kN/m^3. The data are crossing a number of the lines of constant specific solid weight, till it approximates the specific weight of 27 kN/m^3, asymptotely. It appears that all the peat and clayey-peat soils have porosity between 75% and 90%. Peat with porosity lower than 70% is not found.

The axiom $\Delta V=\Delta e$ or $\Delta V=\Delta n$ means that the solid mass is incompressible and the volume change of the soil will be equal to the volume changes of the pores. Therefore, according to this axiom compression of the soil should take place according the red arrows: a decrease of the void ratio and porosity at a constant specific weight of the solids. However, while the organic materials are extremely compressible, Figure 5 shows no peat or clayey-peat with a porosity smaller than about n=70%. Because for the peat and clayey-peat soils this relationship follows the porosity-lines, it means that $\Delta e=0$ and $\Delta V=\Delta V_{solid}/(1-n)$. Therefore, the axiom $\Delta V=\Delta e$ is not valid in general.

Conclusions

The conclusion mentioned above calls for further research. It will not only be necessary to look at the compressibility of the soil itself but also at the compressive behaviour of solid mass and its interaction to the pore pressure and effective stresses. Another important item is the stress on the solid mass directly. This solid stress is caused by the pore pressures as well as by the effective stresses. It may give rise to load dependent secondary settlements, which occur immediately after the first effective stresses occur. Besides the inaccuracy of the method to fit results of the oedometer test, the variation of compression coefficients, the variation of effective weight and consolidation coefficients cause additional significant deviation in prediction. This is especially the case for organic soils. The axiom of the Terzaghi's theory, namely incompressible solids, cannot be applied for organic soils.

References

1. *Masterplan Grondwerken en Funderingen.* CROW rapport 99-1, 2000.
2. Terzaghi, K.: *Erdbaumechanik.* Franz Deuticke, Vienna, 1925.
3. Keverling Buisman, A.S.: Results of long duration settlement tests. *Proc. 1st Int. Conf. Soil Mech. and Found. Eng.*, Cambridge, Mass. No F-7, (1936), pp.103-106.
4. Koppejan, A.W.: A formula combining the Terzaghi load compression relationship and the Buisman secular time effect. *Proc. 1st Int. Conf. Soil Mech. and Found. Eng.*, Rotterdam, 3, (1948), pp.103-106.
5. Taylor, D.W.: *Research on consolidation of clays.* Mass. Inst. Techn. Publication from Dept. Civil & Sanitary Eng., Serial 82, August, 1942, 147p.

Vertical Drains Simulated as Leakage

HANS B. SELLMEIJER
GeoDelft & TUDelft, Delft, the Netherlands

Secondary compression or creep is an important aspect in the design of earth structures on top of soft soils. The key question in such design is the development of pore pressures and settlement with time. In the beginning the soil deformations are delayed due to the low hydraulic conductivity of the subsoil. This causes an increase of the pore pressures. Gradually water starts flowing off, excess pore pressures decrease and settlements develop. Generally, the design of earth structures on top of soft soils is a rather one-dimensional problem. GeoDelft has developed a one-dimensional software program MSettle for prediction purposes. A few other dimensional effects, like load distribution, are incorporated by approximation. The combined process of consolidation and creep is progressing very slowly. In order to limit the excess pore pressures and to accelerate the settlement rate, often a system of vertical drains is applied. Such a system may consist of small permeable vertical synthetic strips or massive granular walls containing drainage pipes. The flow to the vertical drains has a significant horizontal component. This makes it difficult to incorporate it directly in the one-dimensional approach of the MSettle program. Of course, a two dimensional extension is feasible. However, this violates the character of the program: simple and fast. Therefore, an attempt is made to express the horizontal flow into a leakage term, which can be incorporated straight forwardly.

Problem description

The creep behaviour is conventionally described as an empirical relation between linear strain and the logarithm of time (Keverling Buisman [1]). A more fundamental approach is a relation between creep strain, and vertical effective stress. Such models are currently available (Leroueil et al. [3], Edil and Den Haan [4], Den Haan [5,7], Yin and Graham [6], Den Haan and Sellmeijer [9]). These models interrelate strain and effective stress, depending on the creep strain rate. In turn, the effective stress is related to the pore pressure by the equilibrium. Elimination of the effective stress provides a relation between the (volumetric) strain and the pore pressure. This relation is highly non-linear. Strain is recorded as the change of the volume ratio v/v_0, where v_0 is the initial volume. To determine the settlement and pore pressures with time, one extra condition is needed: conservation of mass. This condition reads

$$\frac{\partial}{\partial z}\left(k_z \frac{v_0}{v}\frac{\partial \varphi}{\partial z}\right)+\frac{\partial}{\partial x}\left(k_x \frac{v_0}{v}\frac{\partial \varphi}{\partial x}\right)=\frac{\partial}{\partial t}\left(\frac{v}{v_0}\right)+\frac{v-(1-n)v_0}{v_0 K_w}\frac{\partial w}{\partial t} \quad (1)$$

The notation is the following: φ head, w pore pressure, v volume, k permeability, x, y co-ordinates, t elapsed time, K_w bulk modulus of water, and n initial porosity. The equation is set up for plane flow. A similar equation can be formulated for radial flow.

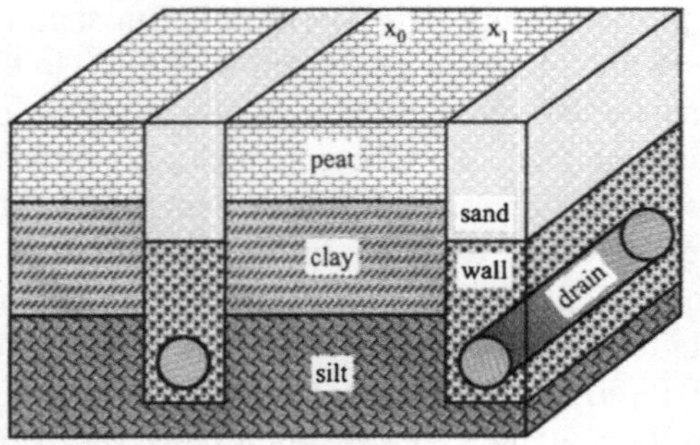

Figure 1. Schematization of the drainage system.

An impression of the plane drainage system is shown in Figure 1. Sand walls are constructed at regular distances. In the walls drain pipes are placed. The pressure in the drain pipes and the air pressure in the sand wall may be adjusted. In the sand wall a water table is formed. Below the table the head is almost constant. Above the table the head equals the vertical position minus the air-pressure head. The horizontal position at the side of the sand wall is indicated by $x = x_0$ and the position between two drains by $x = x_1$. This method of drainage is described for instance in Van Weele [8]. To avoid the need for a two dimensional numerical coding including deviatoric behaviour, the storage equation will be averaged in horizontal direction. If this is feasible, the designer is not bothered to take care of a proper mesh definition and only volumetric strain is involved. The calculation becomes much more simple.

Setting up the horizontal leakage

It is the intention to find an average expression for the storage equation. In fact the volume will already have an averaged value, since the earth structure on top is much stiffer than the underlying soft soil. The loading actually consists of a deformation condition rather than a pressure condition. It is convenient to express the storage equation in shortcut form. The pore pressure is related to the head. We define two linear shortcuts

$$
\begin{aligned}
\mathrm{O}(\varphi) &= \frac{\partial}{\partial\ z}\left(k_z\ \frac{v_0}{\overline{v}}\ \frac{\partial\varphi}{\partial z}\right) - \frac{\overline{v}-(1-n)v_0}{v_0\,K_w}\ \frac{\partial w}{\partial t} \\
\overline{v}_t - \mathrm{O}(\varphi) = \frac{\partial}{\partial x}\left(k_x\ \frac{v_0}{\overline{v}}\ \frac{\partial\varphi}{\partial x}\right) \qquad & \\
\overline{v}_t &= \frac{\partial}{\partial t}\left(\frac{\overline{v}}{v_0}\right)
\end{aligned} \tag{2}
$$

Note that the strain is written in an average notation. We may directly average the storage equation between the vertical drain at $x = x_0$ and the position between two drains at $x = x_1$. The discharge at $x = x_0$ is indicated by Q and the discharge at $x = x_1$ vanishes. The following is obtained

$$
\overline{v}_t - O(\overline{\varphi}) = \frac{v_0}{\overline{v}}\frac{Q}{x_1 - x_0} \qquad \overline{\varphi} = \frac{1}{x_1 - x_0}\int_{x_0}^{x_1} \varphi d\,x \tag{3}
$$

If the leakage occurs from aquitards, the discharge is expressed in the head drop over these layers. Moreover, it is assumed that the average head in the aquifer more or less expresses the head at the interface of aquitard and aquifer. Thus a useful averaged flow equation is obtained (see Verruijt [2]). Here, the procedure to specify a suitable value for the discharge is less obvious. The head at the drain is significantly lower than the average one in the subsoil. The excess pore pressures are intended to dissipate as quickly as possible. Therefore, the behaviour of the head in horizontal direction is crucial to try to find an expression for the discharge. A useful expression of the horizontal behaviour may be obtained by integration of the original storage equation. Three times integration with respect to x must be performed. The first integration yields the horizontal discharge, the second one the head, and the third one the average head. Integration of the storage equation yields

$$
k_x \frac{v_0}{\overline{v}}\frac{\partial\varphi}{\partial \rightleftarrows x} = \int_{x_0}^{x} \{\overline{v}_t - O(\varphi)\}\,d\,x' - \frac{v_0}{\overline{v}}Q \tag{4}
$$

Note that the boundary conditions of the discharge are satisfied. Integration of the discharge yields

$$
k_x \frac{v_0}{\overline{v}}(\varphi - \varphi_0) = \int_{x_0}^{x} d\,x' \int_{x_0}^{x'} \{\overline{v}_t - O(\varphi)\}\,d\,x'' - \frac{v_0}{\overline{v}}Q(x - x_0) \tag{5}
$$

The head at the drain is indicated by φ_0. Averaging of the head yields

$$
k_x \frac{v_0}{\overline{v}}(\overline{\varphi} - \varphi_0) = \frac{1}{x_1 - x_0}\int_{x_0}^{x_1} d\,x' \int_{x_0}^{x'} d\,x'' \int_{x_0}^{x''} \{\overline{v}_t - O(\varphi)\}\,d\,x''' - \frac{v_0}{\overline{v}}Q\tfrac{1}{2}(x_1 - x_0) \tag{6}
$$

In this way the discharge is connected to averaged quantities. However, the triple integration must first be properly worked out.

Working out the horizontal leakage

It is tempting to directly replace the integrand $O(\varphi)$ by its averaged value in order to obtain a compact practical expression for the discharge. But it is uncertain if such an approach is sufficiently accurate. We can do better. The triple integral in (6) is divided in an averaged part and the remaining part. The discharge is written explicitly

$$\frac{v_0}{\bar{v}}\frac{Q}{x_1-x_0}=\frac{2}{(x_1-x_0)^3}\int_{x_0}^{x_1}dx'\int_{x_0}^{x'}dx''\int_{x_0}^{x''}\{\bar{v}_t-O(\bar{\varphi})+O(\bar{\varphi})-O(\varphi)\}dx'''-2k_x\frac{v_0}{\bar{v}}\frac{\bar{\varphi}-\varphi_0}{(x_1-x_0)^2} \quad (7)$$

The constant part of the integrand is processed straight forwardly. The operator O is linear. Therefore, operators in the variable part may be joined and the sequence of integration and operation may be switched

$$\frac{v_0}{\bar{v}}\frac{Q}{x_1-x_0}=\tfrac{1}{3}\{\bar{v}_t-O(\bar{\varphi})\}-O\left[\frac{2}{(x_1-x_0)^3}\int_{x_0}^{x_1}dx'\int_{x_0}^{x'}dx''\int_{x_0}^{x''}(\varphi-\bar{\varphi})dx'''\right]-2k_x\frac{v_0}{\bar{v}}\frac{\bar{\varphi}-\varphi_0}{(x_1-x_0)^2} \quad (8)$$

If the integral term is neglected, the behaviour of the head in horizontal direction appears to be parabolic. This can be observed from (5). The parabolic behaviour must start at the value φ_0 at the drain and satisfy the average head. The following expression meets these conditions

$$\varphi-\bar{\varphi}=(\varphi_0-\bar{\varphi})\left[\tfrac{3}{2}\left(\frac{x_1-x_0}{x_1-x_0}\right)^2-3\frac{x_1-x_0}{x_1-x_0}+1\right] \quad (9)$$

If this behaviour is substituted in the remaining integral and worked out, the discharge is expressed in averaged quantities only

$$\frac{v_0}{\bar{v}}\frac{Q}{x_1-x_0}=\tfrac{1}{3}\{\bar{v}_t-O(\bar{\varphi})\}+\tfrac{2}{15}O(\bar{\varphi}-\varphi_0)-2k_x\frac{v_0}{\bar{v}}\frac{\bar{\varphi}-\varphi_0}{(x_1-x_0)^2} \quad (10)$$

It appears that the discharge is proportional to the average head drop between subsoil and drain. However, there is a correction proportional to flow and storage terms. This correction accounts for the fact that the gradient at the drain is bigger than the averaged head drop per drain distance. Substitution of the derived Q and rearrangement of the result completes the average storage equation (3)

$$\bar{v}_t-O(\bar{\varphi})-\tfrac{1}{5}O(\bar{\varphi}-\varphi_0)+3k_x\frac{v_0}{\bar{v}}\frac{\bar{\varphi}-\varphi_0}{(x_1-x_0)^2}=0 \quad (11)$$

This formulation facilitates one-dimensional calculation. The influence of the drain is translated into a leakage length. The operators are specified in equation (2).

Evaluation

To illustrate the performance of the horizontal leakage, a practical example is presented.. Three layers of peat, clay and silt, respectively, have all a thickness of 5 m. On top of this stratification, located at +0 m, a permeable sand layer is present with a water head at +4 m. At the bottom is a permeable sand layer with a water head at +7 m. The loading consists of three steps. One at 0 days, the next one at 10 days, and the last one at 100 days.

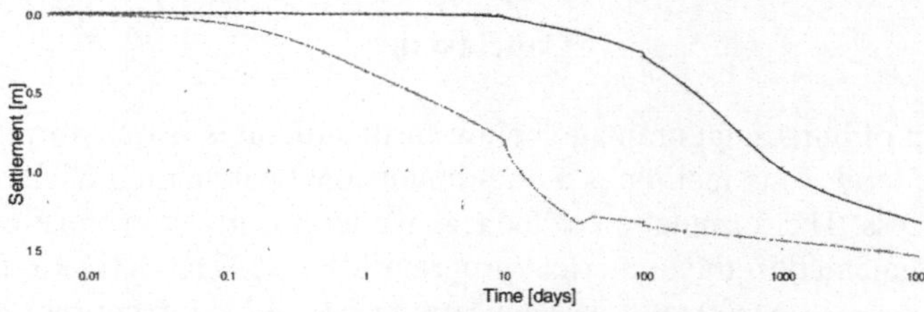

Figure 2. Settlements generated with and without drains

Two calculations are made. In the first one the sand walls are switched off, in the second one they are switched on. The walls have a centre to centre distance of 3 m. They are 13 m long. The peat and clay are fully drained. The silt is partially drained over the top 3 m. During 30 days the water table in the sand walls is lowered to –5 m. After 30 days the water table in the walls is equal to the head in the top sand layer at +4 m.
In the case that the drain is switched off, the settlements are indicated by the pink line in Figure 2. During the first 10 days the settlement rate is very small. After the second load at 10 days the rate increases. After the third load at 100 days again an increase of the rate is observed. In the case that the drain is switched on, the settlements are indicated by the blue line in Figure 2. During the first 30 days the settlement rate is accelerated significantly. When pumping is stopped at 30 days, unloading is observed, followed by a reloading. If no drains are applied, the water dissipates slowly to the top and bottom. Large excess pore pressures may be generated. The drain is forcing the opposite. Large negative excess pore pressures may develop. To give an idea of the size of the pore pressures, they are shown in Figure 3 at two moments: 30 days and 10,000 days.

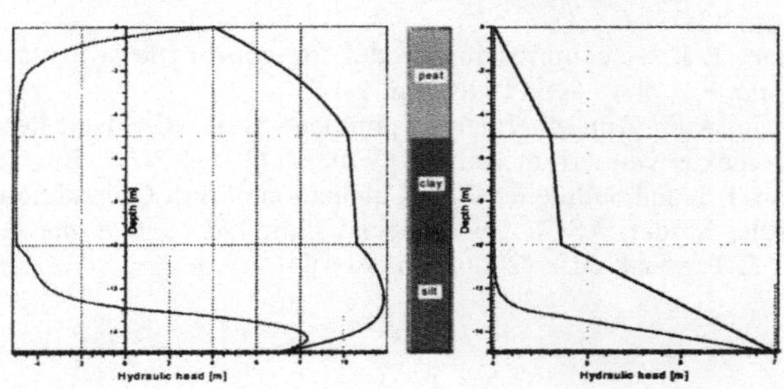

Figure 3. Pore pressures at 30 and 10,000 days

In this example, the normal heads range between +4 and +7 m. After a very long time, when the excess pressures are dissipated, they indeed do, as may be observed by the pink line in the right chart of Figure 3. But in the left chart of Figure 3 the pink line runs up to almost +12 m. However, the drainage tries to force a head of –5 m during the first 30 days. After 30 days the head in the drain is +4 m. In Figure 3 the blue lines clearly show this tendency over the top 13 m.

Conclusion

The effect of horizontal drainage below earth structures is transformed into a leakage term. This facilitates a one-dimensional calculation of the settlement process. The formulation of the leakage term is given in equation (11). It is implemented in the numerical program MSettle. The derivation of the leakage term is performed as accurate as possible. Of course, certain simplifications are introduced as to make the problem suitable for mathematical analysis. The major assumption is an averaged settlement pattern induced by the relative stiffness of the earth structure on top. The leakage approach turns out to have potential. Quite versatile problems may be calculated at a minimum effort.

References

1. Keverling Buisman, A.S.: Results of long duration settlements tests. Proc. 1st Int. Conf. Soil Mech. & Found. Eng., Cambridge, Mass. No. F-7 (1936), pp. 103-106.
2. Verruijt, A.: *Theory of groundwater flow*. MacMillan & Co ltd., London, 1970.
3. Leroueil, S., Kabbaj, M., Tavenas, F. and Bouchard, R.: Stress-strain-strainrate relation for the compressibility of sensitive natural clays. *Géotechnique*, Vol 35(2), (1985), pp. 159-180.
4. Edil, T.B. and Den Haan, E.J.: Settlement of peats and organic soils. ASCE. Geot. Special Publ. 4 (2), (1994). pp. 1543-1572.
5. Den Haan, E.J.: *Vertical compression of soils*. Ph.D. thesis, Delft University Press, Delft, Netherlands, 1994.
6. Yin, J.H. and Graham, J.H.: Equivalent times and one-dimensional elastic visco-plastic modelling of time-dependent behaviour of clays. Can. Geot. J., Vol 31, (1994), pp. 42-52.
7. Den Haan, E.J.: A compression model for non-brittle soft clays and peat. *Géotechnique*, Vol 46 (1), (1996), pp. 1-16.
8. Van Weele, A.F.: Zandschermen en pompen - werkwijze voor het versneld en bouwrijp maken van terreinen. Land+Water 6 (1999), VNU Business Publ.
9. Den Haan, E.J. and Sellmeijer, J.B.: Calculation of Soft Ground Settlement with an Isotache Model. ASCE Geot Special Publ 112, *Soft Ground Technology* (Hanson & Termaat, ed.), (2000), pp. 94-104.

Finite Strain Theory Applied to Gassy Mud Consolidation

BERNADETTE G.H.M. WICHMAN
Ministry of Transport, Public Works and Water Management, Road and Hydraulic Engineering Division, Delft, The Netherlands

In The Netherlands contaminated mud is stored in large scale disposal sites. Because of space limitations, these sites have a depth of order of 50 meters. Natural sediments under anaerobic conditions produce methane gas, which can accumulate in significant amounts in the stored mud layer. It is still uncertain how much gas will be produced on long time scales and also how much gas can escape. Especially at great depth the amount of gas may be large enough to inhibit further consolidation. As a consequence, the storage capacity of the site may be reduced significantly.

During the past 15 years, in the Slufter disposal site, the gas contents as well as the degree of consolidation, have been monitored. It was found that the degree of consolidation was still low with an maximum effective stress level of order of 10 kPa. In addition, significant amounts of gas were found, up to 15% at atmospheric pressure. In the past, the consolidation behaviour of this site has been modelled by means of the model SLIB, based on a modified Terzaghi theory. The agreement was poor, especially because of the presence of undissolved gas. A finite strain consolidation model FSCONGAS for gassy mud has been developed. In addition, suitable laboratory tests were performed on gassy mud, in order to verify the assumptions that have been made [1]. Finally, the new model was used to simulate the consolidation in the Slufter disposal site [1]. It was found that the FSCONGAS model can successfully be applied to predict the disposal capacity. However, there is still a large uncertainty around the amounts of accumulated gas that can be expected on long terms.

Theoretical background

Since the width of the disposal site is usually much larger than its depth, predictions of self weight consolidation can be made by means of one dimensional finite strain theory [2], which has been extended for a gassy soil [3]. The theory that is described here is applicable for soft soils with a gas content below a certain critical value, estimated around 10% for the Slufter disposal site. Among others, the following assumptions were made:

Gas accumulates in large gas bubbles or voids that are much larger than the average pore size of the surrounding soil. This assumption is consistent with scanning electron images of naturally gassy seabed soil samples [4].

Large bubbles are assumed to be fixed inside the soil skeleton, i.e. the bubbles will not escape.

Gassy soil can be idealised as a two phase system, consisting of compressible gas bubbles that are completely embedded in a saturated soil matrix. The consolidation behaviour is described in terms of these two phases, using mechanical concepts for the saturated soil matrix similar to the purely saturated case [4]. In addition, it is assumed that the gas pressure is equal to the total vertical stress plus the atmospheric pressure, with the saturated soil surrounding the gas voids acting to transmit the total stress change to the gas. This is a first approximation derived from oedometer tests [1].

If the self weight consolidation of a gassy sludge is compared to that of a saturated sludge containing the same mass of water and solids, the bulk density is reduced, which gives a smaller driving force for consolidation and a smaller initial excess pore water pressure. In addition, the drainage path length is increased. These effects retard the consolidation process and reduce the final degree of consolidation. The existing finite strain computer program named FSCONBAG has been modified to include this theory.

Field measurements at the Slufter disposal site

Field data from the Slufter disposal site have been used to validate the computer calculations with the FSCONGAS model. The design of the Slufter site comprises a pit with a depth of 29 m below sea level, a surface at sea level of $163*10^4$ m^2 and around a dike with a height of 24 m. The bottom consists of drained sand. Two fixed measuring poles, called North and South, are available for monitoring purposes (see Fig. 1).

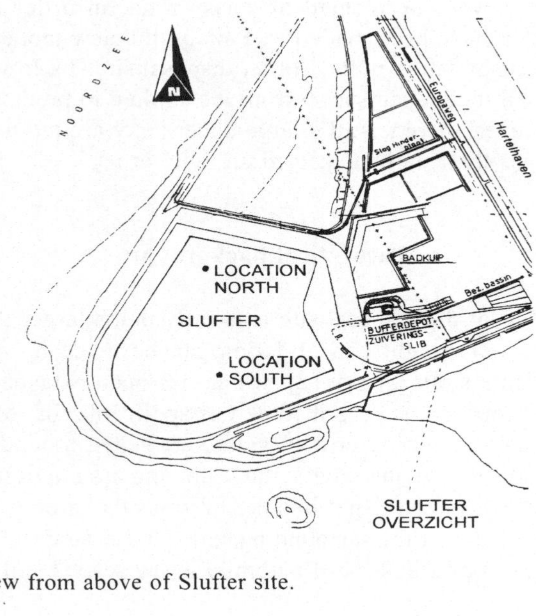

Figure 1. View from above of Slufter site.

In 1994 the sludge level was still below sea level and the water level was approximately equal to the sea level. The mud is pumped to the Slufter site through a long pipeline and is deposited by means of a diffusor. The density in the diffusor varies between 1.03 and 1.28 g/cc, with a mean density of 1.16 g/cc [5]. The main part of the mud is deposited by means of more or less horizontal fluid mud flow. As the position of the diffusor is changed regularly, the depot is filled in a more or less uniform way.

The harbour authorities supplied data about the volume of stored mud from September 1987 until December 1994. The monitoring data consist of 2 monthly echo soundings and 4 monthly backscatter measurements for the determination of the mud level. At the measuring poles North and South profiles of water content, gas content, pore water pressure, temperature and classification properties are available. The samples were taken at 1 m or 1.5 m intervals by means of a special mud sampler, specially developed by Rotterdam Public Works.

Bulk density profiles were calculated from the water contents and the in-situ gas contents, as the backscatter measurement was severely disturbed by the presence of gas. The difference between total stress and pore water pressure was found to be smaller than 10 kPa. This implies that the consolidation is still in an early stage.

The bottom level is likely to be at 29 ± 1 m below sea level. The top level can be found from the deviation of the pore water pressure from hydrostatic. In addition, the top sample indicates the position of the mud surface. The differences in height between both levels are significant, i.e. at maximum 4 meters (see Fig. 2, at location North). In addition, the height was calculated from the nuclear density profiles, temperature profiles, backscatter profiles and echo soundings (see Fig. 2).

Computer calculation: input variables

As the computer program is one dimensional, the geometry of the site has been simplified by superimposing 3 rectangular blocks of mud, with 3 distinct deposition speeds. The deposition speed indicates the accumulation of saturated material and is an average for the whole Slufter site. The gas is added later, during the consolidation calculation.

Consolidation parameters have been determined by means of so called HYDCON-tests [6] and GASCON-tests [1] at GeoDelft. The HYDCON-tests and GASCON-tests were performed on mixed samples of different locations and depth. It was found that the correlations between operative stress, permeability and fluid void ratio are quite similar, from which it was concluded that calculations with parameter sets from GASCON-tests SLU1 and SLU2 are representative for the whole Slufter site. The classification properties were quite similar as well.

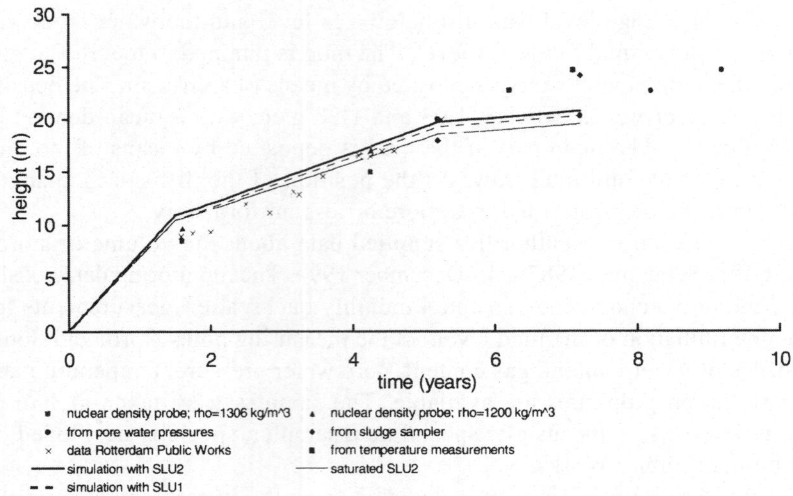

Figure 2. Comparison of measured and simulated height.

Figure 3 shows the gas content at atmospheric pressure as a function of depth for location South in 1992 and 1994. The results for location North are similar. The values vary around 10% to 11%, which indicates that on average the gas content did not increase. For the calculations it was assumed that the initial rise in gas content occurs instantaneously and that the gas void ratio (i.e. gas volume divided by solids volume) at atmospheric pressure is constant.

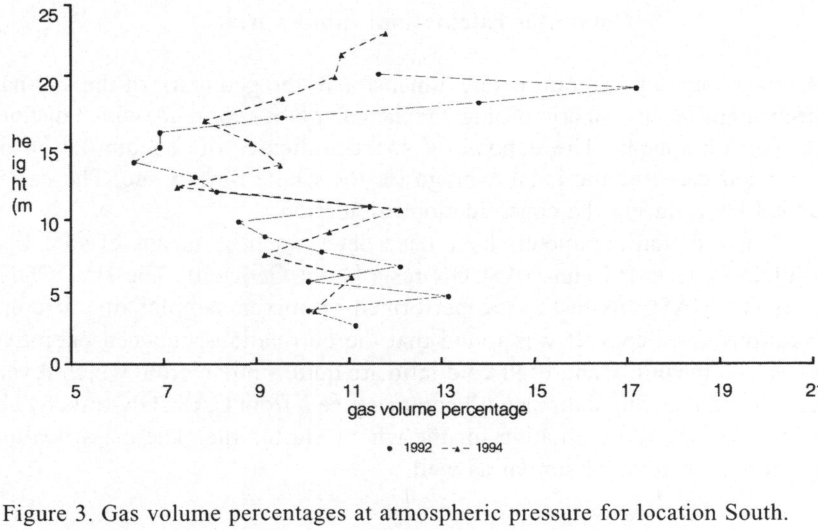

Figure 3. Gas volume percentages at atmospheric pressure for location South.

Computer calculation results

Unsaturated simulations with parameter sets from GASCON-tests SLU1 and SLU2 and a saturated simulation with set SLU2 are shown in Figure 2 for location North starting in September 1987. It is visible that the gassy simulations with SLU1 and SLU2 are reasonably close to the measurements, until 7 years. Possibly, more mud was deposited than the global deposition rate indicates. As all lines are relatively close, it follows that the effect of gas on the mud height is limited, which is consistent with the average measured in-situ gas content around 4%. For location South similar results were obtained. The agreement in 1994 was better in this case.

Figure 4 shows the simulated and measured bulk densities for location South in 1992 and in 1994.

(a) in 1992 (b) in 1994

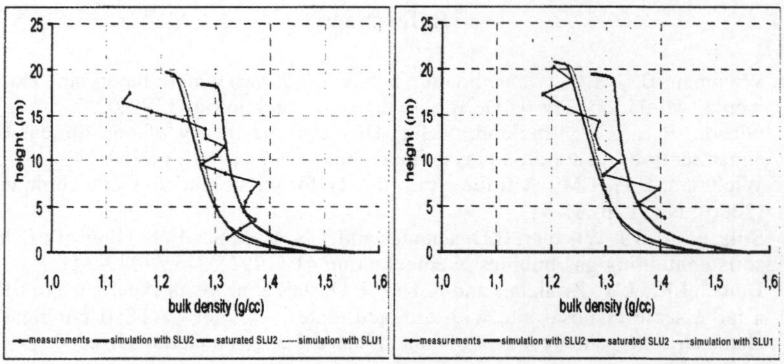

Figure 4. Comparison of measured and simulated bulk density for location South.

It is visible that the unsaturated calculations are in reasonable agreement with the measurements, whereas the agreement with the saturated calculation is less. In the bottom region the measured density is clearly larger than the simulated density in 1992 and in 1994. This due to a higher sand content, which causes a faster consolidation. In 1992 the density in the top layer is clearly smaller than simulated. This is caused by the larger gas content (see Fig. 3), which causes a slower consolidation. Similar effects were locally observed for location North. As the excess pore pressures are concerned, all simulations, including the saturated one, fit the experimental data reasonably well. The differences between the simulations are of the order of the scatter in the experimental data. The effect of adding the gas in the calculations is clearly visible.

Conclusions

The aim was to improve the agreement between the model prediction and the results from the field measurements, by adding the effects of undissolved gas in the finite strain theory, on which the model is based. Indeed, adding the gas in the calculation leads to a significant improvement, which is clear from the bulk density profiles. It is visible that the consolidation is retarded due to the presence of the undissolved gas voids. As the disposal site is still to be filled with another 20 meter of mud, the current findings can be applied for monitoring and the prediction of future consolidation behaviour. Current research is focused on making predictions of the amounts of undissolved gas that can be expected to accumulate in the mud layer.

References

1. Wichman, B.G.H.M.: Consolidation behaviour of gassy mud: theory and experimental validation. Thesis Delft University of Technology (1999).
2. Gibson, R.E., G.L. England, M.J.L. Hussey: The theory of one dimensional consolidation of saturated clays. Géotechnique 17 (1967), p: 261-273.
3. Wichman, B.G.H.M.: A finite strain theory for gassy sludge. Géotechnique 50 (2000), No. 1, p: 35-41.
4. Sills, G.C., S.J. Wheeler, S.D. Thomas and T.N. Gardner: Behaviour of offshore soils containing gas bubbles. Géotechnique 41 (1991), No.2, 227-241.
5. Deibel, I.K., J.W. Zwakhals and A.Th.P.J. Opstal: Nine years experience in filling a large scale disposal site with dredged material. Proc. CATSIII congress on characterisation and treatment of contaminated dredged material, Oostende (1996), p: 219-231.
6. van Essen, H.M. and Greeuw, G.: Combination of laboratory tests to determine consolidation parameter functions for sludge. In Compression and consolidation of clayey soils. Edited by H. Yoshikuni and O. Kusakabe. A.A. Balkema, Rotterdam, The Netherlands (1995), pp. 593-596.

CALCULATION MODELS

Soil Penetration, Numerical Modeling and Application to Microtunneling

PETER VAN DEN BERG
GeoDelft & TUDelft, Delft, the Netherlands

Penetration of rigid devices into the soil is a very complicated mechanical problem. Soil initially located in front of the penetrating device is pushed aside to make room for the device itself. In order to arrive at a steady state solution, a large displacement analysis is necessary.

A promising approach for simulating real steady-state penetration is the strain path method (Baligh [1]). Large-strain updated-Langrangean finite element models are also frequently used to simulate penetration of rigid devices into the soil (Kiousis et al. [2]; Cividini and Gioda [3]). However the updated Lagrangean method has some disadvantages. If very large local deformations occur, elements may distort too much or even turn inside out. The problems due to element distortion can be avoided by uncoupling of nodal point displacements and material displacements. This approach is called an Arbitrary Langrangean Eulerean (ALE) formulation of the problem (Huétink et al. [4]). A special version of an ALE-approach is a full Eulerean approach, in which the finite element mesh is fixed in space, while the soil flows through the elements. The elaboration of such a model and the application to vertical penetration into homogeneous soil is described by Van den Berg [5] and to penetration into layered soil bij Van den Berg et al. [6].

A microtunnel with a diameter of 34 cm was installed at a depth of about 7-meter. An extensive monitoring program was performed in order to investigate the influence on the surrounding soil. Vertical and horizontal soil movements have been measured inside the soil mass and at the soil surface and they are compared to calculation.

Finite element model

In an Arbitrary Langrangean Eulerean (ALE)-formulation, nodal point displacements and material displacements are uncoupled. This implies that convection has to be taken into account in order to be able to update stresses and strains at the nodal points. A method to calculate this convection has been presented by Huétink et al. [4]. Then additional continuous stress and strain fields are introduced by extrapolating and interpolating nodal point stresses and strains from integration point values. The convective terms are calculated as products of gradients of the additional continuous fields and the displacement increments. This will be illustrated for a simple one-dimensional example. The velocity of a material particle (x^m) at a spatial location x at time t is defined as v^m.

$$v^m = \frac{dx^m}{dt} \tag{1}$$

The velocity of a nodal point (x^n) is

$$v^n = \frac{dx^n}{dt} \tag{2}$$

The material rate of change of a stress measure at a spatial location x at time t can be written as follows

$$\dot{\sigma}^m = \left[\frac{d\sigma}{dt}\right]^m = \frac{\partial\sigma}{\partial t} + v^m \frac{\partial\sigma}{\partial x} \tag{3}$$

And the rate of change of the same stress measure in a nodal point can be expressed as

$$\dot{\sigma}^n = \left[\frac{d\sigma}{dt}\right]^n = \frac{\partial\sigma}{\partial t} + v^n \frac{\partial\sigma}{\partial x} \tag{4}$$

Eliminating the spatial rate of the stress from equations. (3) and (4) yields

$$\dot{\sigma}^n = \dot{\sigma}^m + \left(v^n - v^m\right)\frac{\partial\sigma}{\partial x} \tag{5}$$

Using a fixed mesh, i.e. all nodal points are fixed in space, equations (5) reduces to

$$\dot{\sigma}^n = \dot{\sigma}^m - v^m \frac{\partial\sigma}{\partial x} \tag{6}$$

So, the stress s in a nodal point with a fixed location x after a time increment Δt can be written as follows

$$\sigma^n(x,t+\Delta t)=\sigma^n(x,t)+\dot{\sigma}^m\,\Delta t-\left(v^m\,\Delta t\right)\frac{\partial\sigma}{\partial x} \tag{7}$$

in which the third term is the additional convective term. To get a stable solution some numerical diffusion is required (Huétink et al. [7]). The consequence of the addition of the convective term is that material including its stresses and strains 'streams' through the fixed finite element mesh.

Numerical analysis of microtunneling

In order to install small-diameter pipelines in density populated urban areas microtunneling techniques can be used. In Japan those techniques are becoming more and more popular and are frequently applied for the installation of telecommunication or electricity cables and gas pipes. Many different types of installation systems can be used. In soft clayey soils, fully soil-displacing techniques are available. The application of soil displacing systems causes surface heave during and after installation. The amount of surface heave depends on the diameter, the soil conditions and the depth of the microtunnel. The Eulerean finite element model has been applied to the simulation of this microtunneling process in layered soil. A fully soil-displacing microtunnel with a diameter of 34 cm is horizontally pushed into the soil at a depth of about 6.8-meter.

The finite element discretization consists of 8000 8-noded three-dimensional cubic elements with a selective reduced integration scheme (i.e. one-point integration for the volumetric part of the behaviour and eight-point integration for the shearing part, Nagtegaal et al., [8]). Since a lubrication fluid has been applied between tunnel and soil, the interface behaviour has been assumed to be fully smooth. Around the microtunneling machine, however, friction between machine and soil has been included. The loading consists of a material flow through the (fixed) mesh, introduced by prescribed displacements at the side of the mesh in front of the microtunnel face.

The soil consists of three different layers. To model the constitutive behaviour of the soil, a linear-elastic–perfect-plastic. Von Mises model has been applied. This implies that the behaviour of the soil is assumed to be fully undrained. The material parameters are summarised in Table 1.

Table 1. Material parameters.

	Depth		E (MPa)	υ (-)	C_u (kPa)
	from	to			
Layer 1	0.0 m	-3.8 m	11.2	0.49	12.1
Layer 2	-3.8 m	-4.8 m	56.0	0.49	30.2
Layer 3	-4.8 m	-10.2 m	4.0	0.49	34.7

The calculation has been carried out using a tangential stiffness approach, and a global force unbalance criterion equal to 0.1%. The calculation was stopped after a horizontal penetration distance of 16.5 meter. At that point a steady state solution was obtained.
Displacements of the surrounding soil are very important for the acceptance of microtunneling techniques in urban areas. For that reason the results presented in this paper are mainly focussed on soil movements.

Results and comparison to monitoring results

The calculated results indicate that the horizontal displacements around the tunnel are limited to an area of about 1.5-meter around the tunnel. The maximum heave occurs at a distance of about 7 meters behind the tunnel front (steady state condition). The maximum heave for this specific situation is 3.3 mm. A view of the calculated heave at the soil surface is given in Figure 1.

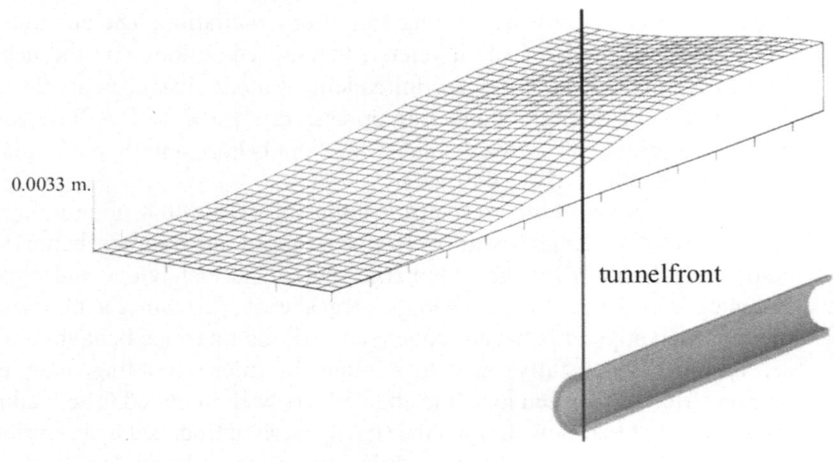

Figure 1. Calculated heave due to microtunneling (steady state solution)

Now, the results of the calculation are compared to monitoring results, carried out recently on Awaji-island near Osaka, Japan, during the installation of a microtunnel. In Figure 2 the heave at ground level is compared to measured data. The maximum value of the heave at ground level is very small, less than 3.5 mm according to both the calculated and measured results. The accuracy of the monitoring devices is about 1 – 1.5 mm.

In Figure 3 the calculation results of the horizontal displacements in z-direction are compared to monitoring data. According to both calculated and measured results this area is restricted to approximately one meter above and one meter below the tunnel. The calculation results are in close agreement to the measurement data, both qualitatively and quantitatively.

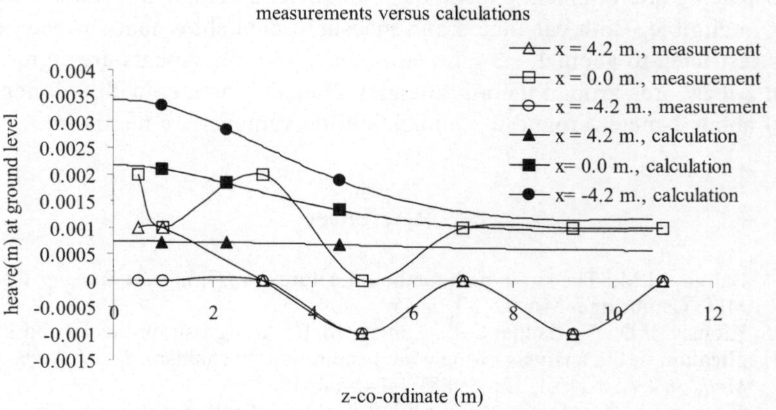

Figure 2. Comparison between calculated and measured heave at ground level.

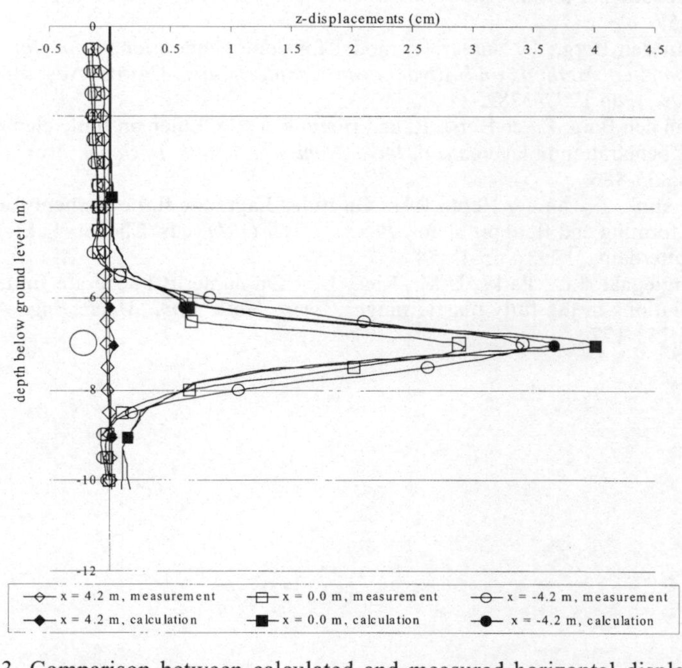

Figure 3. Comparison between calculated and measured horizontal displacement around microtunnel.

Conclusion

The application of a 3-dimensional Eulerean finite element model yields reliable results both with respect to the numerical procedure and with respect to monitoring data. The results indicate that surface heave due to fully soil displacing microtunneling techniques (diameter 34 cm, depth about 7 meter) is negligible. Both calculated and measured data show that surface heave is restricted to about 1 – 3 mm. Horizontal soil movements are significant in a local area around the microtunnel. Outside this area with a dimension of about 2-meter around the tunnel, soil movements are negligible.

References

1. Baligh, M.M.: The strain path method. Lecture *1, MIT Special Summer Course*, MIT, Cambridge, MA 02139, 1985.
2. Kiousis, P.D., Voyiadjis, G.Z., Tumay, M.T.: A large strain theory and its application in the analysis of the cone penetration mechanism. *Int. J. Num. Anal. Meth. in Geom.*, Vol. 12, (1988), pp.45-60, 1988.
3. Cividini, A., Gioda, G.: A simplified analysis of pile penetration. *Proc. of 6th Int. Conf. on Num. Meth. in Geom.*, (1988), pp. 1043-1049.
4. Huétink, J., Vreede, P.T., van der Lugt, J.: Eulerean-Langrangean finite element simulation of forming processes. *Int. J: Num. Meth. Eng.*, Vol. 30, (1990), pp.1441-1457.
5. Van den Berg,: P. Numerical model for cone penetration. *Proc. Int. Conf. on Computer Methods and Advances in Geomechanics*. Cairns, Australia, Vol. 3, (1991), pp.1777-1782.
6. Van den Berg, P., de Borst, R. and Huétink, J.: An Eulerean finite element model for penetration in layered soil. *Int. J. Num. Anal. Meth. in Geom.*, Vol. 20, (1996), pp.865-886.
7. Huétink, J., van der Helm, P.N.: On Euler-Lagrange finite element formulation in forming and fluid problems. *Proc. NUMIFORM* (eds. Chenot et al.), Balkema: Rotterdam, (1992), pp.45-54.
7. Nagtegaal, J.C., Parks, D.M., Rice, J.C.: On numerical accurate finite element solutions in the fully plastic range. *Comp. Meth. Appl. Mech. Eng.*, 4, (1974), pp.153-177.

The Changing Role of Computer Models in Geotechnical Engineering and Design

RONALD B.J. BRINKGREVE
TUDelft & PLAXIS bv, Delft, the Netherlands

In the past, computer models have played a significant role in geotechnical engineering and design and will continue to do this in the future. However, the type of computer models and the level of application have changed over the years. Considering the fact that computer models have mainly evolved from science, it is logical to observe that science has always been ahead of engineering in the use of computer models. Since science and education are closely related (at least at universities) new developments in computer modeling may find their way to the engineering practice through the education system, but this is not the only way. Institutes working in between science and engineering play an important role in bridging the gap between the two worlds.

This presentation illustrates the development and application of computer models in geotechnical engineering and design, especially from a 'Delft' viewpoint. It also demonstrates the role of science and education in computer model development and the task of institutes in making models readily available for the engineering practice. In the final chapter an outlook is given into the future.

Computer models

Computer models are numerical calculation procedures implemented in a computer program and meant for the solution of a specific physical process (or a particular problem). Examples of conventional computer models for geotechnical engineering and design are Winkler type (or subgrade reaction) models for the calculation of displacements and forces of plates in or on the ground, and the method of slices for the calculation of a global stability factor for embankments or slopes. In addition to these relatively simple models, advanced computer models based on the finite element method exist for the solution of boundary value problems, such as deformations or groundwater flow, either two-dimensional (2D) or three-dimensional (3D). Computer models are somehow used in geotechnical engineering to analyse existing or future situations under certain conditions or to design constructions, taking into account intermediate construction phases.

Past

Calculation procedures, numerical methods and constitutive models for geotechnical engineering and design existed well before 1960, but these models were only used in hand calculations since computers were not available. A nice example of such a model is Blum's method for the design of retaining walls. In engineering, the main issue was stability, whereas deformation got less attention.

After the introduction of the third generation computers and computer programming standards around 1964 it became possible to implement models in computer programs and to mount these on mainframe computers using punch cards. The 1960's was also the period in which a new family of non-linear soil models was developed in the UK (Cam-Clay). In the same period, the finite element method evolved, but the method was initially only applicable for linear processes.

In the 1970's various computer models for mainframe computers were developed by researchers. Attention shifted from linear calculations to non-linear solution procedures, which was essential for the use in geotechnical applications. In Delft, the Dutch centre of civil engineering, quite a few non-linear finite element programs were developed of which only two examples are given here: DIANA, a TNO program, initially only meant for concrete structures, and ELPLAST, a Delft University program for geotechnical applications, later evolved as PLAXIS. These models were mainly used for research purposes related to prestigious projects, such as the Eastern Scheld Storm Surge Barrier. In the same period Rijkswaterstaat Bouwdienst in co-operation with LGM (at present GeoDelft) built a spring model for the design of retaining walls, called DAMWAND. This model was mainly used by researchers and young engineers who actually became computer modeling experts.

Shortly after the introduction of home and personal computers in the early 1980's, enthusiastic engineers in various companies started to make simple computer programs to speed up calculations that used to be done by hand. This resulted in several individual PC programs in which simple computer models were implemented. In that time, worked, a.o., on a finite element model for groundwater flow as a replacement for the electrical analogon. In the mid 1980's, Frans Barends at GeoDelft initiated the development of a series of DOS-based PC programs, the Mseries for geotechnical engineering and design, including a spring model for retaining wall design (MSHEET, evolved from DAMWAND), a stability program based on Bishop's method (MSTAB), a settlement analysis program based on Koppejan's formula for one-dimensional compression (MZET) and a groundwater-flow program (MSEEP). This set the standard for geotechnical engineering and design in late 1980's. At the same time, Arnold Verruijt in the Geotechnical Lab of Delft University created similar programs and stimulated the use of computer models by students and civil engineers. As a result, a new generation of civil engineers became familiar with the use of conventional computer models in geotechnical engineering and design.

A sub-group of researchers in the Geotechnical Lab, led by Pieter Vermeer, was working on the first PC version of PLAXIS. The initial idea was to create an advanced but practical and user-friendly tool for the analysis of river dikes, to be used by geotechnical engineers of Rijkswaterstaat (DWW). The range of applications increased when structural elements were added, but the philosophy did not change. At the end of the 1980's, the program was ready for geotechnical engineering in general, but it took some time before the engineering practice accepted the finite element method and learned how to determine the parameters of the Mohr-Coulomb model.

Many scientists were sceptical about the idea that engineers without numerical backgrounds could use the finite element method. They focused on the modeling of complex phenomena like dynamics, liquefaction, softening and their numerical implementation in three-dimensional multi-purpose finite element programs on fast UNIX work stations, implicitly increasing the distance between science and engineering. Also DIANA evolved in this way, but the PLAXIS group persisted in their philosophy and continued their developments in view of practical geotechnical finite element applications.

In the early 1990's, the benefits of the finite element method convinced more geotechnical engineers. They still used conventional methods for *design*, but increasingly they used the finite element method for a detailed *analysis* of large projects, since deformation became equally important as stability.

Half way the 1990's, when it became business policy to concentrate on core competence, there was the transition from DOS to Windows. This forced most engineering firms that maintained their own PC programs to stop such activities. Not all programs could make the transition, since quite an investment was required to adapt the user-interface to the complex Windows system. The transition was also an opportunity for program developers who did succeed, to make programs more versatile and user-friendly, which was appreciated by practical users. Apart from user-friendliness, more advanced soil models became available. Before these models were understood and appreciated by the users, quite some time passed.

Over the years, various companies had assembled large databases with geotechnical records, but most companies used a different system to store their data. The use of GIS for object-oriented. data and statistics and probabilistic gained interest. However, most of these activities were separate from each other. The exchange of data between GIS, databases and computer models was quite limited. A new standard had to be set.

Present

At the turn of the century, the performance of personal computers had increased exponentially, so that it became possible to perform 3D non-linear finite element calculations on powerful PC's. On the other hand, the increasing complexity of civil engineering projects required robust and easy-to-use 3D computer models. What had been a privilege for science has become feasible for the engineering practice. However, in order to maintain a low entrance level of 3D models, computer programs should focus on particular applications rather than becoming multi-purpose. This idea resulted in the release of 3D 'specials', such as the DIANA Borehole Special, created by TNO, and the PLAXIS 3D Tunnel program.

Unless the need for 3D models, the well accepted 2D models and even the conventional methods are still being used in the engineering practice and they all have their specific role. Design is generally based on conventional methods. At the start of a project, when only limited soil data is available, a quick finite element calculation is generally done with a simple soil model to get qualitative insight in the variety of stress paths and to determine the locations where more soil investigation is required. As a second step, advanced model parameters are determined from the additional soil investigation data and a more detailed finite element analysis is done with advanced soil models to quantify deformations, structural forces and to determine the influence of the project on existing structures. 3D finite element models are mainly applied for large and complex projects.

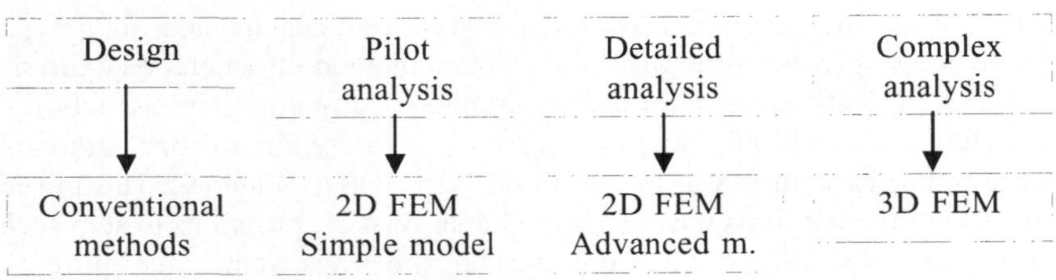

Figure 1. The present role of computer models in a project

In education, students are taught in the possibilities and limitations of advanced computer models and the meaning of model parameters than morethey are educated in how to implement these methods in computer programs. This is a logical trend, since most students are going to work in the engineering practice where a sound working knowledge of computer models is preferred over detailed scientific backgrounds, as long as limitations of the models are understood. The number of students entering research and contributing to the development of computer models has dramatically reduced. Moreover, it is recognized that more attention should be paid in education on all aspects of model parameter determination.

In engineering practice, companies have realised that it is beneficial to store geotechnical data in a standard format. This has resulted in the GEF standard for the exchange of geotechnical data. This is only a start. There are initiatives for an integrated use of ICT in civil engineering projects in general, such as VISI (CUR). There are also initiatives to provide computer models, data and services over Internet (DelftSystems).

Future

In the new millennium the need for application-oriented 3D finite element models will increase. Projects become even more complex, but also more international due to the process of globalisation. This requires calculation methods that are independent from local standards; i.e. methods in which reality is realistically simulated using advanced constitutive models. Such models will have more than a few parameters. Although these parameters must have a proper anc accepted meaning, there is a need for facilities that support the user in Soil Schematization and determining model parameters while giving clear insight in the consequences of a particular set of parameters. Also the variation of parameters in a probabilistic context becomes a real issue. Safety analyses will gradually be replaced by failure probability analyses. Advanced computer models must be able to deal with variations of loads, parameters and geometric dimensions and to search for the optimum or most critical situation in an automatic way, while stimulating the responsibility and creativity of the engineer. All together, it will make advanced computer models being more and more used as a tool for *design* rather than just for *analysis*, whereas the use of conventional methods will gradually decrease

This new way of design in civil engineering projects involves a large amount of data, which should be available through databases and additional soil investigation. Facilities and international standards have to be created for the exchange of data between testing equipment, databases, GIS and computer models, and for the elaboration of data from databases (a.o. parameter determination). Compatibility transparency and transmissibility of data are keywords for the future.

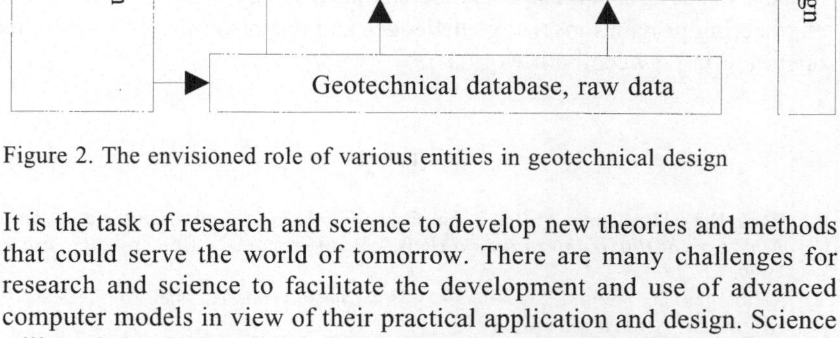

Figure 2. The envisioned role of various entities in geotechnical design

It is the task of research and science to develop new theories and methods that could serve the world of tomorrow. There are many challenges for research and science to facilitate the development and use of advanced computer models in view of their practical application and design. Science will (and should) be ahead of the engineering practice, but it is important that both worlds interact with each. On one hand, the research world should realise the problems and needs of the engineering practice. On the other hand, the engineering practice should be open to new theories and models that improve the quality of engineering and design. Nevertheless, the distance between both worlds will remain and cannot just be overcome through the education system. It also gives the right of existence to institutes (either assigned by the government, or on a commercial basis), which are dedicated to bridge the gap between science and engineering, a.o. by making robust and user-friendly computer models and by providing post-educational training and support. The key of success of such institutes is that they are creative and flexible, that they understand the world of science and the world of engineering and that they are aware of the large distance in between both worlds.

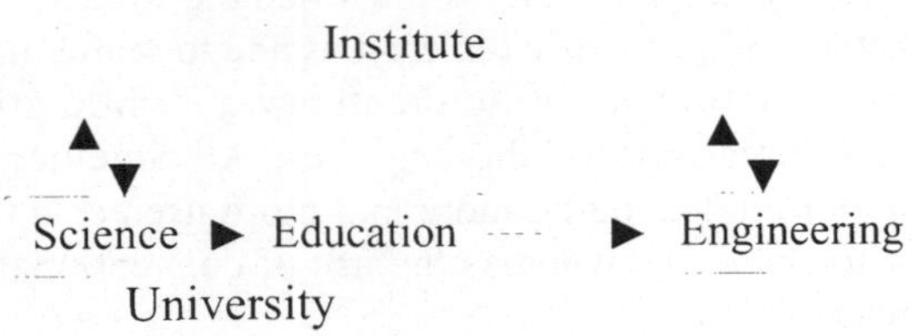

Figure 3. The relationship between science and engineering

Conclusion

The complexity and application of computer models in geotechnical engineering and design has changed over the years. Nowadays, geotechnical engineers are familiar with conventional and advanced methods, although the latter are used for *analysis* more than for *design*. A further automation of data transfer and elaboration has to be achieved, including facilities for parameter determination, variation and optimisation. In this process there are many challenges for the research world in scientifically developing new theories and methods to facilitate the evolution and use of advanced computer models for *design* purposes. The persisting distance between science and engineering provides inspiring challenges and responsibilities for institutes operating in between both worlds.

References

1. Burd, H.J.: The history of Plaxis. In: *Beyond 2000 in Computational Geotechnics Ten Years of Plaxis International* (R.B.J. Brinkgreve, ed). Balkema, Rotterdam 1999.
2. Rietbergen, D., Steijn, L.: *Computers - Moderne slaven in een nieuwe tijd. Samsom* Uitgeverij, Alphen a/d Rijn 1972.
3. Verruijt, A.: *Grondmechanica*. Delft University Press,1999.

Finite Element Solution Coupled Consolidation Equations

MASSIMILIANO FERRONATO

Dept. Mathematical Methods and Models for Scientific Applications, University of Padova, Italy

The consolidation theory was developed in a three-dimensional (3D) setting by Biot [1], giving rise to a system of PDEs that combines the elastic equilibrium of the porous body with the continuity of ground water flow.

Because of the mathematical complexity of Biot's consolidation theory, analytical solutions are possible only for simple problems with a symmetric configuration, e.g. [2], while for more complex realistic situations it is necessary to use numerical techniques. At present the implementation of a finite element (FE) consolidation model can be considered a quite common practice. Nevertheless, the solution to realistic problems may still be very difficult, particularly because the resulting linear system can be ill-conditioned, especially for small time steps, and typically has a large size, thus requiring significant computational cost and effort.

The present paper addresses some difficulties which arise from the FE solution to the coupled consolidation equations and suggests efficient numerical techniques for solving large size realistic models. A criterion for the critical time step leading to ill-conditioning is developed and compared with the accuracy condition by Vermeer and Verruijt [3]. Finally, the use of iterative methods is explored, with a special attention to reordering, preconditioning, and scaling strategies which can accelerate the solver convergence.

Theory overview

The distribution of stresses and pore pressures in a saturated porous body during consolidation is governed by the principle of effective stress, equilibrium, and continuity of flow. Equilibrium is prescribed by applying the virtual work principle to the whole system and using the effective stress definition, while the continuity of flow is governed by the classical storage equation [2] integrated by the Galerkin method of weighted residuals with a weak formulation.

The geomechanical and flow models are solved simultaneously for the nodal displacements δ and pore pressures p:

$$\begin{bmatrix} K & -Q \\ 0 & H \end{bmatrix} \begin{Bmatrix} \delta \\ p \end{Bmatrix} + \begin{bmatrix} 0 & 0 \\ Q^{\mathrm{T}} & P \end{bmatrix} \begin{Bmatrix} \dot{\delta} \\ \dot{p} \end{Bmatrix} = \begin{Bmatrix} f^{\mathrm{u}} \\ f^{\mathrm{p}} \end{Bmatrix} \quad \Rightarrow \quad K_1 x + K_2 \dot{x} + f = 0 \qquad (1)$$

where K, H, Q, and P are the elastic and flow stiffness, coupling, and capacity matrices, respectively. Observe that K, H, and P are symmetric and positive definite. Equation (1) is solved in time by an appropriate marching scheme; one of the most popular is the so-called θ-method:

$$[\theta K_1 + K_2/\Delta t]x_{t+\Delta t} = [K_2/\Delta t - (1-\theta)K_1]x_t - [\theta f_{t+\Delta t} + (1-\theta)f_t] \tag{2}$$

Equation (2) constitutes the linear system $A_{\Delta t}x_{t+\Delta t}=b_t$ which is solved for the displacements and pore pressures in time domain. The matrix controlling the numerical consolidation model:

$$A_{\Delta t} = [\theta K_1 + K_2/\Delta t] = \begin{bmatrix} \theta K & -\theta Q \\ Q^T/\Delta t & \theta H + P/\Delta t \end{bmatrix} \tag{3}$$

is non-symmetric and needs to be factorized at each time step because of the Δt variability during the simulation if a direct solution method is used. It can be observed that $A_{\Delta t}$ could be symmetrized by multiplying the second set of equations by $-\theta\Delta t$. However, this gives no practical advantage in the numerical solution to (2) since the resulting matrix has negative eigenvalues. Consequently, the stability of the solution becomes an issue for some values of Δt [4], and the classical Conjugate Gradient method cannot be used. For these reasons, in the following the solution to (2) will be studied using the matrix $A_{\Delta t}$ in the unsymmetric form (3).

Ill-conditioning of the coupled system

In a typical ill-conditioned linear system, two or more rows are nearly parallel. This condition applied to rows i and m of the consolidation matrix (3) leads to:

$$\frac{\theta K_{ij}}{Q^T_{mj}/\Delta t} \cong \frac{-\theta Q_{ik}}{\theta H_{mk} + P_{mk}/\Delta t} \qquad j = 1,\ldots,n_u \quad k = 1,\ldots,n_p \tag{4}$$

where n_u and n_p are the number of unknowns for nodal displacement and pore pressure, respectively. The Δt satisfying equation (4) causes ill-conditioning of $A_{\Delta t}$, and so it will be denoted as the critical time step Δt_{crit} for the problem.

Equation (4) can be solved by substituting the terms of the blocks forming the global $A_{\Delta t}$ with a rough estimate of their magnitude. Let us assume tetrahedral elements for 3D problems and annular elements with a triangular cross-section for axisymmetric problems. The coefficient K_{ij} is proportional to the elements of the elastic stress strain matrix, and hence has the order of magnitude of the Young modulus E. Similarly, H_{mk} and P_{mk} are approximately proportional to k/γ, with k the permeability and γ the water specific weight, and $\Sigma\phi\beta$, respectively, with ϕ the porosity and β the water compressibility. Σ represents a characteristic dimension of the FE grid, which corresponds to the elemental area in case of an axisymmetric problem. The Q terms are the projections of the element boundary along the coordinate directions, so they can be approximated by $\sqrt{\Sigma}$. Replacing the above approximations into (4) and solving for Δt we obtain:

$$\Delta t_{\text{crit}} \cong \left(\frac{1-\psi}{\theta}\right)\frac{\Sigma\gamma}{kE} \tag{5}$$

with $\psi=\phi\beta E$. Note that in the previous procedure the influence of the mesh geometry has been neglected, so equation (5) must be viewed only as the basic relation between Δt_{crit} and the physical parameters of the porous medium through a characteristic measure of the triangulation. More generally, we can write Δt_{crit} as [5]:

$$\Delta t_{\text{crit}} = X(\psi,\theta)\frac{\Sigma\gamma}{kE} \tag{6}$$

where X is an unknown dimensionless factor depending upon ψ, θ, and the shape and resolution of the mesh.

Equation (4) allows for an interesting geometric interpretation. Write equation (4) as follows:

$$\Delta t(\theta K_{ij}/Q_{\text{mj}}^{\text{T}}) \cong -\Delta t\theta_{\text{ik}}/(\Delta t\theta H_{\text{mk}} + P_{\text{mk}}) \Rightarrow R_{\text{l}}(\Delta t) \cong R_{r}(\Delta_{t})$$

and compare graphically the functions R_{l} and R_{r}, as shown in Figure 1.

It can be noted that for $\Delta t > \Delta t_{\text{crit}}$ the difference between R_{l} and R_{r} grows with the time step size, so the matrix parallelism reduces and the system is well-conditioned. By contrast, for $\Delta t = \Delta t_{\text{crit}}$ R_{l} remains close to R_{r}, and consequently the system is expected to be ill-conditioned below the critical condition as well.

Observe in Figure 1 that an intersection between R_{l} and R_{r} exists only if:

$$[\text{d}R_{\text{l}}/\text{d}\Delta t]_{\Delta t=0^{+}} < [\text{d}R_{r}/\text{d}\Delta t]_{\Delta t=0^{+}} \tag{7}$$

which represents a necessary condition for the existence of Δt_{crit}. Equation (7), with the aforementioned approximations for the matrix coefficients, yields:

$$E/\sqrt{\Sigma} < \overline{\psi}\sqrt{\Sigma}/(\Sigma\phi\beta) \Rightarrow \psi < \overline{\psi} \tag{8}$$

where $\overline{\psi}$ is a factor which accounts for the quantities neglected in our estimates, i.e. basically the mesh geometry. Thus, it can be argued from equation (8) that a $\overline{\psi}$ value exists for ψ above which $A_{\Delta t}$ is well-conditioned

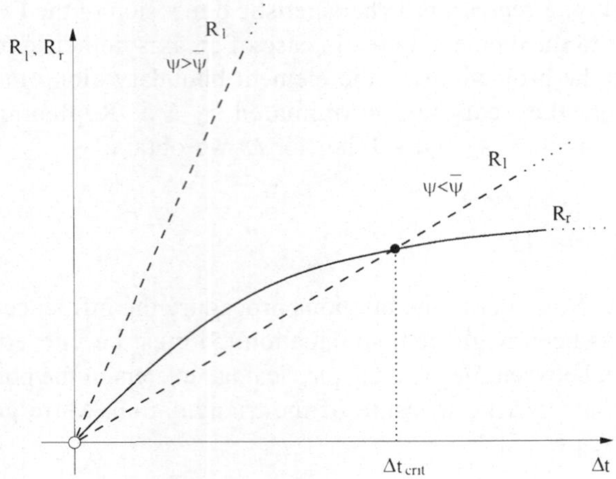

Figure 1. Qualitative behavior of left-hand side R_l and right-hand side R_r of equation (4) vs. Δt and $\psi=\phi\beta E$.

independently on the time step size.

A different lower limit for Δt was derived by Vermeer and Verruijt [3] based on the accuracy of the pore pressure solution in a consolidating porous column. This bound was explicitly calculated for 1D problems as:

$$\Delta t \geq 1/6\,(\Delta h)^2/(\theta c) \tag{9}$$

with c the consolidation coefficient, and Δh a characteristic size of the FE grid. The authors claim that a similar expression holds true for 2D and 3D problems as well. Replacing into equation (9) the expressions

$$c = k/[\gamma(\mu/E+\phi\beta)];\ \mu = (1+\nu)(1-2\nu)/(1-\nu);\ (\Delta h)^2 \cong \Sigma,$$

with ν the Poisson ratio, leads to:

$$\Delta t_{crit} = \left(\frac{\mu+\psi}{6\theta}\right)\frac{\Sigma\gamma}{kE} = X'(\psi,\theta)\frac{\Sigma\gamma}{kE} \tag{10}$$

Note that equation (10) is very similar to equation (6), although the implications of the two time step limits are different. This shows that ill conditioning and accuracy of solution are basically connected one to another since they depend on the same combination of physical parameters. The main difference between the critical time steps arising from equations (6) and (10) stems from the weighting factors X and X'. In fact, while for the ill conditioning problem a $\overline{\psi}$ value may be found for ψ above which no Δt_{crit} exists, in the accuracy condition X' is never zero or negative, and hence a lower bound for the time step size always exists.

Solution by Conjugate Gradient like methods

Conjugate Gradient like solvers appear to be very attractive for the solution to large size sparse sets of indefinite equations, since they require a relatively limited amount of core memory and can be easily implemented on parallel computers. One of the most popular and efficient algorithms belonging to this class is the bi-conjugate gradient stabilized (Bi-CGSTAB [6]), which proves to be a robust and cost-effective solver for consolidation problems [4] if appropriate preliminary numerical strategies are adopted in order to mitigate the ill-conditioning of the FE coupled equations.

The natural way of numbering the variables suggested by (1) and (3), with the displacements coming first and pore pressures coming next, gives rise to a very large bandwidth. An inexpensive and much more convenient way of ordering the unknowns originating a reduced bandwidth is to nest displacements and pore pressures, so that the vector $\boldsymbol{x}_{t+\Delta t}$ of (2) is made of the components $(\delta_{x1}, \delta_{y1}, \delta_{z1}, p_1, \delta_{x2}, \delta_{y2}, \delta_{z2}, p_2, \ldots)$.

To be competitive with direct and partitioned solvers, Bi-CGSTAB must be effectively preconditioned. A popular preconditioner is derived from the incomplete triangular factorization of $A_{\Delta t}$ with zero fill-in (ILU(0)). However, in ill-conditioned problems ILU(0) may provide a poor approximation of $A_{\Delta t}$ [5], thus resulting in a lack of Bi-CGSTAB acceleration. A recent evolution of ILU(0) is ILUT(ρ,τ) [7], which allows for a partial fill-in of the triangular factors controlled by the parameters ρ (number of newly created elements preserved in each row) and τ (bound on the magnitude of the preconditioner coefficients).

Since one source of ill-conditioning of FE consolidation equations stems from the large difference in size between the terms of the geomechanical and of the flow equations, the use of appropriate preliminary scaling strategies may also prove an important operation to reduce the spectral condition number of $A_{\Delta t}$. The most intuitive scaling is the Standard Jacobi algorithm, where each row/column is scaled by the diagonal term. A more sophisticated algorithm is the one by Curtis and Reid [8], prescribing that the scaled matrix has its entries near to unity in the sense that the sum of the squares of the logarithms of its coefficients is minimized. An implementation of this algorithm is provided by the routine MC29 of the Harwell Software Library 2000.

Preconditioned Bi-CGSTAB is experimented with preliminarily scaled and reordered FE consolidation equations of a realistic numerical example. An axisymmetric porous volume 4500 m large and 3000 m deep, consisting of alternating sandy (k=10^{-5} m/s) and clayey (k=10^{-8} m/s) layers with E=$8.33\cdot10^6$ kg/m^2 and ν=0.25, is pumped from an aquifer located at a depth of 1000 m. The FE system totalizes 65703 unknowns and is integrated in time using Δt=100 s, which is found to be very close to the critical time step from the spectral analysis of $A_{\Delta t}$ [5]. The convergence profiles are reported in Figure 2 and the CPU times in Table 1. Note that the use of an efficient scaling strategy and an improved preconditioner allows for a significant acceleration of the convergence rate and a reduction of the

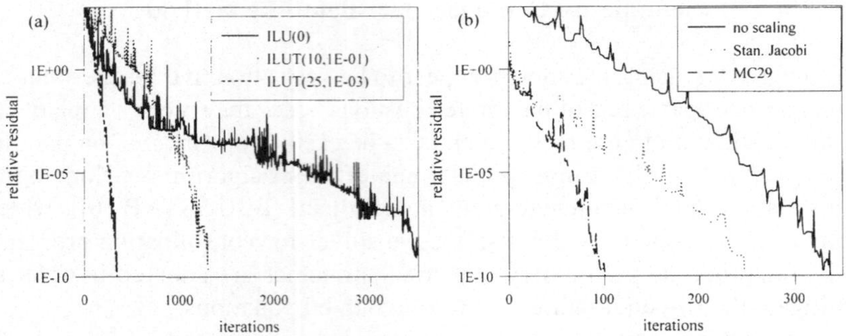

Figure 2. Bi-CGSTAB convergence profiles: (a) with no scaling and different preconditioners; (b) ILUT($25,10^{-3}$) preconditioner and different scaling strategies.

computational cost.

It is worth emphasizing that the computational performance obtained in a realistic, large size, and severely ill-conditioned problem is very efficient. So the technique suggested in the present paper appears to be a promising solution approach to FE coupled consolidation models.

Table 1. CPU times (s) for the test-case system on a 600 MHz workstation.

	scaling	preconditioner	convergence	total
ILU(0) without scaling	—	0.60	504.00	504.60
ILUT($25,10^{-3}$) without scaling	—	6.69	78.10	84.79
ILUT($25,10^{-3}$) + Stan. Jacobi	0.02	6.69	65.57	72.28
ILUT($25,10^{-3}$) + MC29	0.59	10.33	27.80	38.73

References

1. Biot, M.A.: General theory of three-dimensional consolidation. J. Appl. Phys. 12 (1941), p:155-164.
2. Verruijt, A.: Elastic storage of aquifers. In R. De Wiest (ed.): Flow Through Porous Media. Academic Press, New York, 1969, pp.331-376.
3. Vermeer, P.A. & Verruijt, A.: An accuracy condition for consolidation by finite elements. Int. J. Numer. Anal. Meth. Geomech. 5 (1981), p:1-14.
4. Gambolati, G., Pini, G. & Ferronato, M.: Numerical performance of projection methods in finite element consolidation models. Int. J. Numer. Anal. Meth. Geomech. 25 (2001), p:1429-1447.
5. Ferronato, M., Gambolati, G. & Teatini, P.: Ill-conditioning of finite element poroelasticity equations. Int. J. Solids Struct. 38 (2001), p:5995-6014.
6. Van der Vorst, H.A.: Bi-CGSTAB: a fast and smoothly converging variant of Bi-CG for the solution of nonsymmetric linear systems. SIAM J. Sci. Stat. Comp. 13 (1992), p:631-644.
7. Saad, Y.: ILUT: a dual threshold incomplete ILU factorization. Num. Lin. Alg. Appl. 1 (1994), p:387-402.
8. Curtis, A.R. & Reid, J.K.: On the automatic scaling of matrices for Gaussian elimination. J. Inst. Math. Applic. 10 (1972), p:118-124

The Practical Value of Slope Stability Models

ANDRÉ R. KOELEWIJN
GeoDelft, Delft, the Netherlands

In the past century a large number of models to analyse the stability of slopes have been proposed, first by Fellenius [1,2], shortly followed by others, for instance May and Brahtz [3] and Bishop [4]. As all models are only an idealization of reality, one should be careful to trust the results. Common idealizations for slope stability analysis were the adoption of a simplified, usually circular failure surface and the assumption of plane strain conditions. In spite of these simplifications, a slope stability analysis by hand was quite laborious, as a sufficient number of trial surfaces need to be examined to find the most critical one. This has become a lot easier by the development of computer programs.

Increased computational power has enabled engineers to concentrate more on obtaining the right input for a slope stability analysis and on selecting the appropriate model. The latter is thought to be particularly relevant in special cases where more commonly used methods, for example Bishop's method or a finite element model, do not represent the failure mode well enough or are likely to diverge. The problem of obtaining representative input data is of a more general nature, as in practice the heterogeneity of the subsoil usually plays an important role.

In this article the influence of the stability model commonly used is compared to the influence of an individual geotechnical consultant involved and uncertainties regarding the soil profile and the mechanical properties of the soil layers. The comparison is made for a situation where the uplift phenomenon is important.

The uplift phenomenon

Due to high water pressures in a sand layer below soft layers beneath and behind a dike, uplift induced slope failure may occur. Failure is caused by reduction of the shear stresses at the interface between the sand layer and the soft layers due to high water pressures in this sand layer, which are in turn caused by a high water level in the river, see Figure 1. The uplift phenomenon mainly occurs in tidal river basins. Case histories of uplift induced dike failure are, for instance, known from the river Lek (the Netherlands) [5] and the river Thames (England) [6].

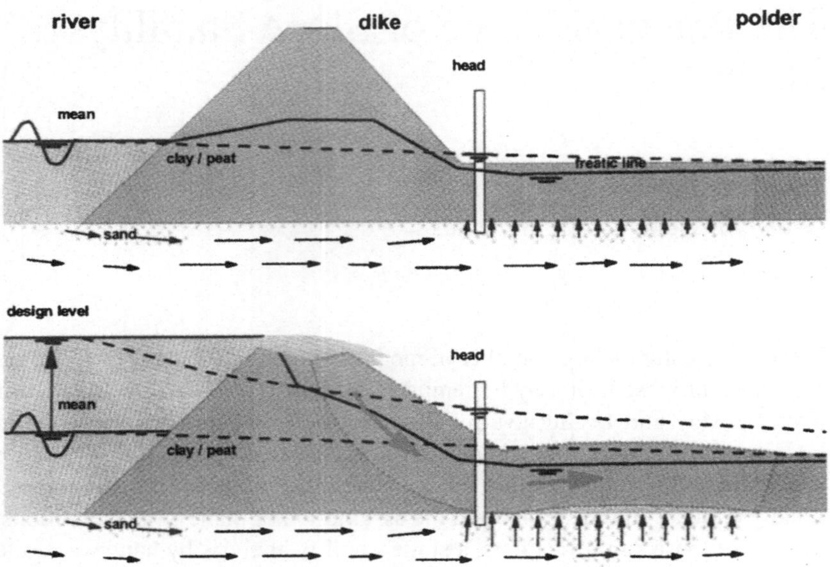

Figure 1. Schematic process of uplift induced slope failure (from [5]).

Dutch dike design and evaluation practice

The uplift failure mechanism has been included in the Dutch national guideline for dike design and evaluation [7]. As a part of a large research project aimed at the improvement of this guideline, five geotechnical consultants have been asked to evaluate the slope stability of a small part of the river dike near Bergambacht, located about 30 kilometres east of Rotterdam along the river Lek and protecting the Krimpenerwaard polder. The information initially provided to these consultants was similar to the information usually available for the systematic evaluation of Dutch primary river dikes, carried out every five years. At stage I, general information regarding the design level, a plan, some cross-sections and two cone penetration test results were supplied. During stages II and III of the research, more detailed information such as regionally representative shear strength properties of different soil types, local piezometric recordings, a boring and results from a third cone penetration test were provided.

The results of the first stage varied widely. The calculated stability factor ranged from 0.52 to 0.84, and the schematization of the soil profile by the consultants differed significantly. The schematizations given in Figure 2 and Figure 3, for instance, are based upon exactly the same information. In both cases, Van's method [5] indicated a lower stability factor than Bishop's method. According to some other consultants, the outcome of Bishop's method was critical, instead.

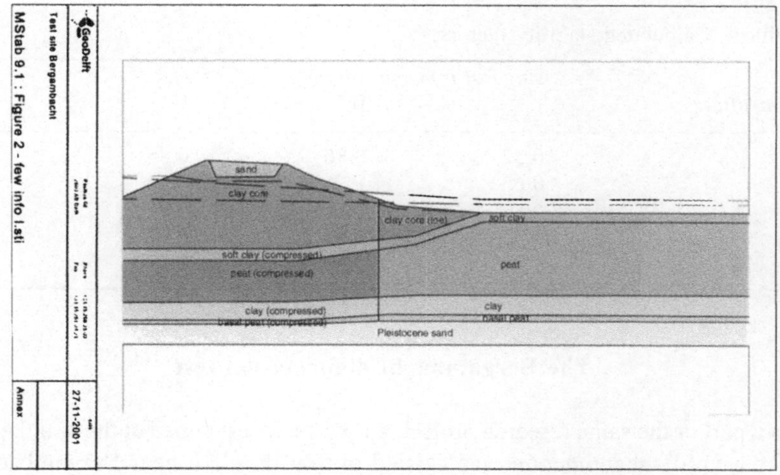

Figure 2. Soil profile based upon two cone penetration tests (I)

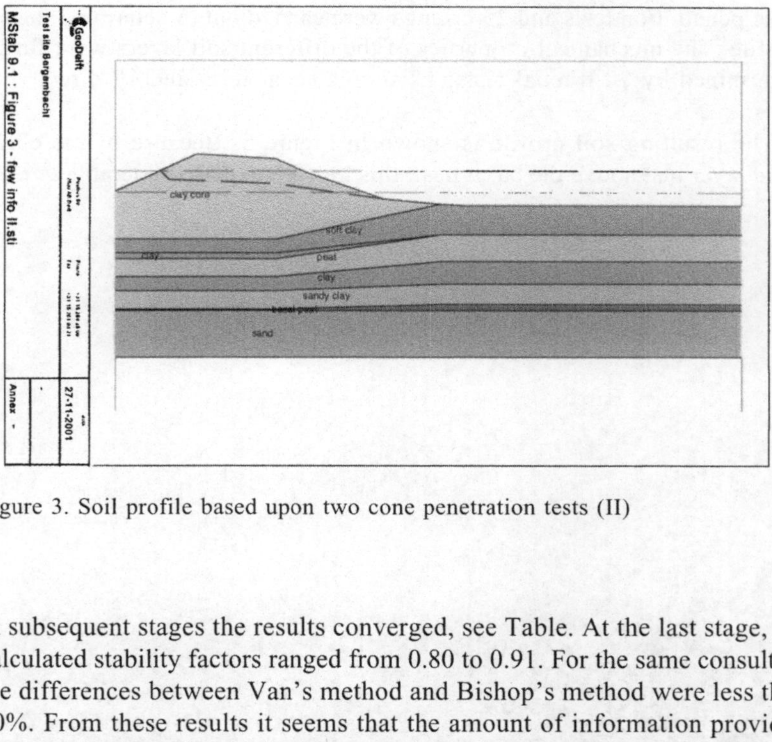

Figure 3. Soil profile based upon two cone penetration tests (II)

In subsequent stages the results converged, see Table. At the last stage, the calculated stability factors ranged from 0.80 to 0.91. For the same consultant the differences between Van's method and Bishop's method were less than 10%. From these results it seems that the amount of information provided at stage I is insufficient for a proper determination of the stability factor and thus of the safety of a dike. The differences between the individual consultants indicate that the introduction of a 'consultant-factor' as an analogue of the model factor should be considered.

Table 1. Calculated stability factors.

	stage of information		
Consultant	I	II	III
1	0.52	0.86	0.85
2	0.63	-	-
3	0.84	0.82	0.80
4	0.75	0.92	0.91
5	-	0.82	0.87

The Bergambacht deformation test

As a part of the same research project, a field scale test aimed at the simulation of the uplift phenomenon was carried out on the dike near Bergambacht. In Figure 4 an aerial view of the test site is given. Behind the existing dike a new dike has been built to enlarge the river bed. The width of the test site, in front of the new dike, is about sixty metres. At the test site over 40 cone penetration tests and 25 borings were carried out to determine the soil profile. The mechanical properties of the different soil layers were further determined by 70 triaxial tests, 11 simple shear tests and 10 direct shear tests.

The resulting soil profile is shown in Figure 5. Because of the clayey sand layer just above the basal peat, this location is less vulnerable to uplift

Figure 4. Test site near Bergambacht

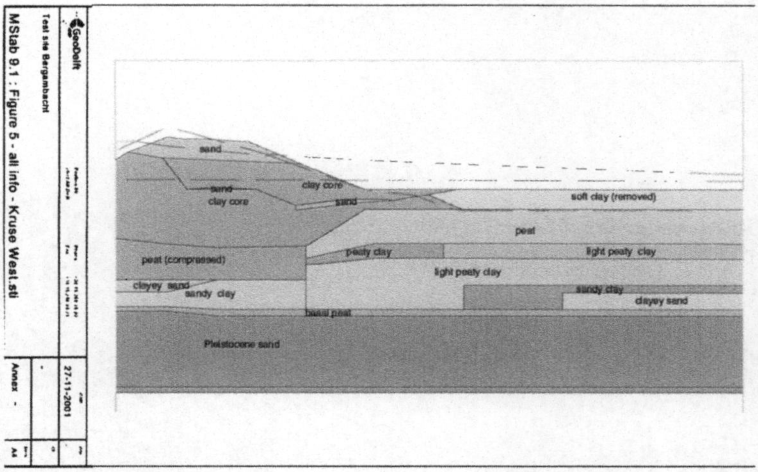

Figure 5. Soil profile at test site based upon all available information

induced failure than most of the neighbouring parts of the river dike. Therefore, the upper two meters of soft clay have been removed, which has resulted in an uplift safety of the polder at the test site comparable to the uplift safety behind those neighbouring parts of the dike.

As the design level of the river is not available on demand, the test is carried out by the infiltration of a large amount of water (up to 1200 cubic metres per hour) into the Pleistocene sand layer, beneath the old dike. Two tests were carried out in Autumn 2001.

During the first test, in September, water was only infiltrated during high tide. After an initialization period, the dike started to move. Beneath the toe of the dike, at the Pleistocene-Holocene interface a distinct shift of 2.5 cm was observed.

During the second test, at the end of November, water was infiltrated during a longer period. After two days the dike collapsed due to uplift. In Figure 6 an overview of the test site after failure is shown. The collapsed soil body is shown in Figure 7. Near the toe of the dike the polder has risen by about 60 cm, while the maximum deformation at the top of the dike, shown in Figure 8, is about 1.5 metres.

Analyses using Van's method, Bishop's method and Plaxis [8] all indicate that the stability factor during the first stage was about 1.0, while during the second stage the predicted minimum stability factor is slightly less than 0.9, using mean values for all relevant soil properties, while disregarding three-dimensional effects.

For the same situation, the difference between Van's method and Bishop's method was generally less than 10%, while the difference between these methods with Plaxis could be up to 30%. For some part, this may be explained by the fact that in Plaxis the full uplift potential could not be applied, as it led to insuperable numerical difficulties.

Figure 6. Test site after failure

Figure 7. Collapsed soil body

Figure 8. Top of the dike after failure

The use of characteristic values for the shear strength of the different soil layers appeared to be much more important than the calculation model used. Using lower bound values, as usual for design, for a certain stage of the test a stability factor as low as 0.36 was calculated, while the calculation using mean values resulted in 1.00 and the use of upper bound values resulted in a value of 1.65 [9].

Conclusions

Apparantly, for the current engineering practice the choice of the slope stability calculation model is not very important when compared to the importance of sufficiently detailed soil investigations, individual expert characterization of the soil profile and uncertainties regarding the shear strength of the soil. For the Bergambacht test site the latter appeared to be the most important factor in determining the slope stability.

Acknowledgement

This study, and particularly the large field test, was supported by Delft Cluster and the Road and Hydraulic Engineering Division of the Dutch Ministry of Public Works.

References

1. Fellenius, W.: *Erdstatische Berechnungen mit Reibung und Kohaesion*. Ernst, Berlin, 1927.
2. Fellenius, W.: Calculation of the Stability of Earth Dams. *Trans. 2nd Cong. Large Dams* (1936), pp.4:445.
3. May, D.R. and Brahtz, J.H.A.: Proposed Methods of Calculating the Stability of Earth Dams. *Trans. 2nd Cong. Large Dams* (1936), pp.4:539.
4. Bishop, A.W.: The use of the slip circle in the stability analysis of slopes. *Géotechnique* 5 (1955), pp.7-17.
5. Van, M.A.: New approach for uplift induced slope failure. *ISSMGE*: Proc. XVth ICSMGE, Istanbul (2001), pp.2285-2288.
6. Cooling, L.F. and Marsland, A.: Soil Mechanics Studies of Failures in the Sea Defence Banks of Essex and Kent. *Inst. Civil Eng.:* Conference on the North Sea Floods of 31st January – 1st February, 1953. Inst. Civil Eng., London (1953), pp.58-73.
7. Technische Adviescommissie voor de Waterkeringen, *Leidraad Toetsen op Veiligheid (Guidelines for the Evaluation of Safety)*. TAW, Den Haag, 1999 [In Dutch].
8. Brinkgreve, R.B.J. and Vermeer, P.A.: *PLAXIS , finite element code for soil and rock analyses*. Balkema, Rotterdam, 1998.
9. Koelewijn, A.R.: *Definitief ontwerp vervormingsproef (Final design of deformation test)*. GeoDelft CO-710301/208v4, Delft, 2001 [In Dutch].

The Initial Stress Problem

KAREL L. MEIJER
MEYER, Engineering & Scientific Services, Giethoorn, the Netherlands

For the analysis and the prediction of the behaviour of soil caused by external load changes, computer models based on the finite element method are often used. Well known packages are, among others, DIANA (De Witte [1]) and PLAXIS (Brinkgreve & Vermeer [2]).

A program for the computation of saturated soil under plane strain conditions was developed in the eighties at WL|Delft Hydraulics in co-operation with GeoDelft and the Technical University Delft. In this computer program, WAZAN, geometrical non-linearity caused by large deformations and/or large displacements is taken into account (Meijer [3]).

For a realistic analysis of soil behaviour it is often necessary to implement complicated relations between occurring stress changes and strains. Numerous more or less sophisticated stress-strain models have been developed, each with its own scope. Nearly all models predict a change of strain as a function of a change of stress depending on the actual stress level in the soil. Here we reach a weak spot of many models: which stress exists at the start of the calculation?

For an initially flat, horizontal ground surface the vertical initial stress in a point below the surface can easily be determined from the weight of the upper layers in the soil column. Assuming that the principal stress axes are uniformly oriented, horizontally and vertically, the horizontal stress can be obtained via an estimated or experimentally determined K_0-value. For more complex configurations one might estimate the initial stresses, but these will generally not satisfy the equilibrium equations.

One method to obtain initial stresses which satisfy the equilibrium equations is making a computation with a linear-elastic stress-strain model. A drawback is then that stress fields may be computed which are not possible in soil. For example in cohesionless soil, tension stresses are not possible, but in a linear-elastic model they are often generated. Thus the simulation of soil subject to a change of load often has a false start: at first the *linearly-elastic* failed soil must be *repaired* with the help of a non-linear (for example an elasto-plastic) soil model. The obtained initial stress may be a surprise.

Another objection against this method is that the origin of the soil mass with its existing stresses is simulated by a stress-strain model which has been developed for total different circumstances.

In this paper a method is given to determine the initial stresses in a different way. The resulting stress field satisfies the equilibrium equations and no stress-strain relation is used. The method uses the infrastructure that exists inside finite element programs and it can be easily incorporated in such programs. The method is described and illustrated on the basis of the WAZAN-program, but it is also applicable in other finite element programs.

The Finite Element Method

The FEM usually applied is the so-called displacement method. In this method a vector with displacements (or displacement increments) is defined in a number of nodes and determined as the solution of a system of equations. For the theory we refer to the numerous textbooks about these matters, for example Zienkiewicz [4]. In This section a summary is given as a basis of the treatment of the initial stress problem.

We consider a plane strain situation and small displacements and deformations. *Stress* is understood to be the total stress, i.e. the sum of the effective stress and the pore water pressure.

The area of interest is divided into triangular elements. The three corner points and the three midpoints of the sides form the nodes in which the displacements are defined. The displacements are interpolated quadratically over the area.

The two equilibrium equations for the two normal stresses in x- and y-direction, σ_x , σ_y and the shear stress τ_{xy} are

$$B^T \sigma = k \tag{1}$$

in which B^T is a differential operator matrix, σ the stress vector and k the vector with body forces respectively

$$\begin{bmatrix} \frac{\partial}{\partial x} & 0 & \frac{\partial}{\partial y} \\ 0 & \frac{\partial}{\partial y} & \frac{\partial}{\partial x} \end{bmatrix}, \begin{bmatrix} \sigma_x \\ \sigma_y \\ \tau_{xy} \end{bmatrix}, \begin{bmatrix} 0 \\ -\gamma \end{bmatrix} \tag{2}$$

The normal stresses for pressure are positive (N/m^2), the positive y-direction is upward, γ is the unit weight of the soil (N/m^3). The stress can be expressed in the strain via the 3x3-matrix D containing the stress-strain relations

$$\sigma = D\varepsilon \tag{3}$$

The strain vector is similarly defined as the stress vector and contains the components ε_x, ε_y en γ_{xy}. The strain vector arises by differentiation of the displacement vector. The latter is approximated element-wise via quadratic interpolation between the six nodes of the element.

The 3x6-matrix with quadratic interpolation functions for element i is called A_i. These matrices are combined in a global matrix A so that

$$\varepsilon = Au \tag{4}$$

u is the $2n_k$-vector with displacements in all nodes.

Applying the Galerkin method in which the equilibrium equations are multiplied with weight functions and integrated over the area, results in a global system of equations for the displacements. The same functions are applied to weight as well as interpolation functions. So we get

$$\int_V A^T B^T D(BA)u \, \mathrm{d}W = \int_V A^T k \, \mathrm{d}W \tag{5}$$

After partial integration of the left hand side there finally arises

$$\{\int_V A^T B^T D(BA) \, \mathrm{d}W\}u = \int_V A^T k \, \mathrm{d}W + \int_S A^T r \, \mathrm{d}s = K_b + K_s \tag{6}$$

The last vector in the right hand side is the contribution of the external stresses integrated along the boundary: the boundary forces. The area integral is computed by summing up the contributions per element which are obtained by a three-point Gaussian quadrature formula. System (6) is reduced by eliminating the rows and columns that correspond with prescribed displacements.

Stress equations

The vector $D(BA)u$ in equation (6) represents the stress vector as a function of the location. The global stress field is determined by the $2n_k$ displacements of the nodes. For a known stress field a vector S can be computed according to

$$S = \int_V (BA)^T s \, \mathrm{d}W \tag{7}$$

For non-linear stress-strain relations (D depends on the stress) equation (6) is solved iteratively. With the obtained stress S can be obtained from equation (7). The residual vector $R = K_b + K_s - S$ is a measure for the deviation of the equilibrium. In finite element programs the residual is iteratively diminished below a certain norm. The final solution is dependent on D. Convergence is not always guaranteed.

Now, we will search for a stress field that obeys the equilibrium equations and for which certain restrictions are prescribed without taking into account the stress and deformation history explicitly. To this aim a discrete stress vector is defined in the side nodes of the elements. These nodes also serve as integration points for the three point integration formula. Inside the element the stress varies linearly.

So with a number of n_z side nodes there are n_z stress vectors with $3n_z$ components to be determined. For a particular choice of these stress vectors the residual vector R that consists of $2n_k$ components can be computed.

Because there are only two equilibrium equations it is assumed that a linear relation exists between the stress components. This can be accomplished by defining a 2-vector q_e of which the three stress components are obtained via a 3x2-matrix G_e , so

$$s = G_e q_e \tag{8}$$

If for example the requirement that the horizontal stress must be K_0 times the vertical stress, then the relation (8) becomes

$$\begin{bmatrix} \sigma_x \\ \sigma_y \\ \tau_{xy} \end{bmatrix} = \begin{bmatrix} K_0 & 0 \\ 0 & 1 \\ 0 & 1 \end{bmatrix} \begin{bmatrix} q_{e1} \\ q_{e2} \end{bmatrix} \tag{9}$$

The parameters q_{e1} and q_{e2} are in this case equal to σ_y and τ_{xy}. Another possibility is to pose $\tau_{xy} = 0$, that means that the principal directions are horizontal and vertical.

The continuous q_e-distribution is approximated by linear interpolation from the $2n_z$ q_e-values collected in the global vector q. The linear interpolation polynomials are collected in the $2x2n_z$ -matrix L. So the residual becomes

$$R = \int_V A^T k \, dW + \int_S A^T r \, ds - \{\int_V (BA)^T GL \; dW\} q \tag{10}$$

By minimization of the L_2-norm of the residual vector R , i.e. $R^T R$, a system of equations for q is obtained. In other words q is determined by the least squares method. If system (10) is shortly written as

$$Mq = K - R \tag{11}$$

then by applying the least squares method, the following arises

$$M^T M q = M^T K \tag{12}$$

If the matrix on the left hand side of equation (6) has a banded structure, then the (symmetrical) matrix $M^T M$ also has it. However, its bandwidth is about three times as large.

The foregoing derivation is valid when the external forces only consist of body forces and known boundary forces. In order to prevent free translation and rotation, at least three displacements must be prescribed. The reaction forces are, as yet, unknown and they depend on the stress strain relations in the case of static indeterminacy. This problem is solved by excluding from the equation (10) the degrees of freedom which correspond with the prescribed displacements. The size of system (12) is not influenced by this. After solving q and consequently σ, the reactions may be calculated via (10).

Example

The method described in the previous section has been applied in the configuration drawn in Figure 1. It concerns the cross section of a dike, that after determination of the initial stress, is subject to an increasing pore water pressure until failure occurs. In this paper we will consider only the computation of the situation at the beginning.

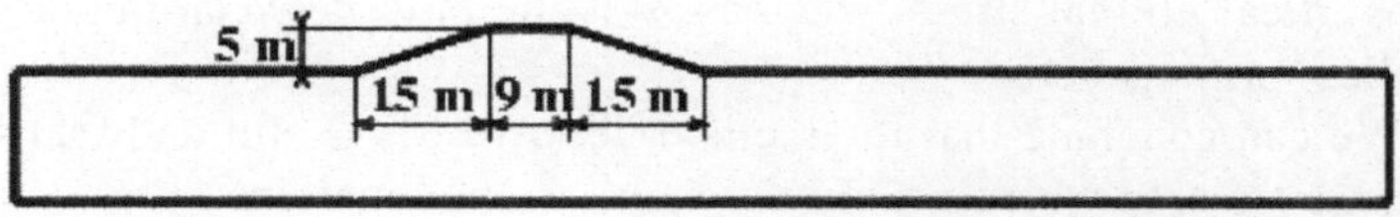

Figure 1. Lay-out of the computational domain.

The number of triangular elements is 740, the number of nodes 1575 and the number of side nodes is 1157. The initial stress is computed with a linear-elastic model. The unit weight of the soil is stated 20 kN/m^3, the bulk modulus 20 MN/m^2 and Poisson's ratio 1/3. This value corresponds with a value for K_0 of 0.5 near the lateral boundaries.

A relevant measure for the stability of cohesionless soil is the ratio between the largest and the smallest principal stresses, this is called the flow factor in Lambe & Whitman [5]. We call the inverse of this quantity W. At failure according to Mohr-Coulomb, W_f is:

$$W_f = \frac{1 - \sin\varphi}{1 + \sin\varphi} \tag{13}$$

Wherein ϕ is the friction angle. For $\phi = 30°$, W_f equals 1/3. In Figure 2 some contours of W are shown.

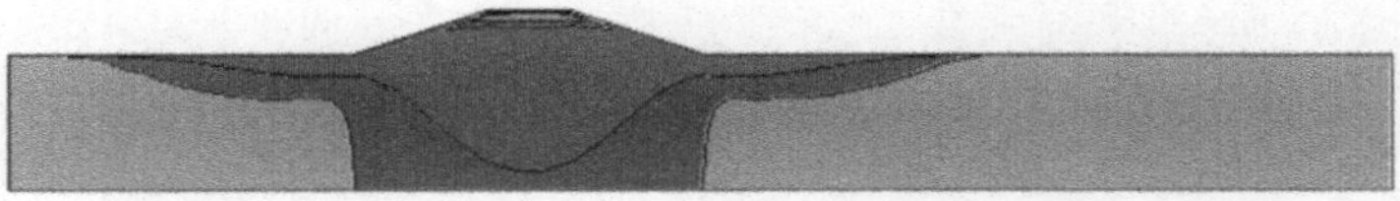

Figure 2. Linear elastic solution. Contours of W, red: $W < 1/3$, green: $1/3 < W < 0.48$, blue: $0.48 < W \leq 0.5$.

From this figure it is clear that the linear-elastic solution is not acceptable for many soil types.

To compute the initial stresses with the method presented, a G-matrix has been used according to equation (9) with $K_0 = 0.5$. The method yields a nearly uniform W-value. In the green parts of Figure 3 W lies inbetween 0.48 and 0.45.

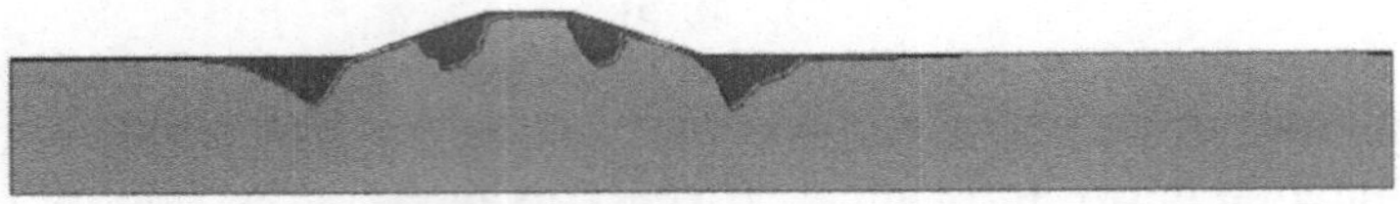

Figure 3. Method presented in this article. Contours of W, green: $0.45 < W < 0.48$, blue: $0.48 < W \leq 0.5$.

The deviations from 0.5 arise because K_0 is the ratio between the horizontal and vertical normal stress, whereas W is the ratio between the largest and smallest principal stress.

We can conclude that the method described is useful and that it yields realistic stresses which satisfy the equilibrium equations.

References

1. Witte, F.C. de., ed.: *DIANA Users Manual*, release 7. TNO Building and Construction Research , Delft , 1998.
2. Brinkgreve, R.B.J. & Vermeer, P.A.: *PLAXIS, Finite Element Code for Soil and Rock Analyses*, version 7, Users Manual. Balkema, Rotterdam 1998.
3. Meijer, K.L.,: *Computation of Stresses and Strains in Saturated Soil*, Doctor's Thesis, TUDelft, 1985.
4. Zienkiewicz, O.C.: *The Finite Element method in Engineering Science*. McGraw-Hill, London 1971.
5. Lambe, T.W. & Whitman, R.V.: *Soil Mechanics*, John Wiley, New York 1969.

Analysis of Piled Raft Foundation by Finite and Infinite Prism Elements

ER-XIANG SONG, HUAI Z. GAO
Tsinghua University, China

Piled raft foundation is widely used for various buildings. The analysis method for this type of foundation has been an important research topic for geotechnical engineers [1][2]. Randolph once gave a review of the available analysis methods [2]. Generally they can be classified into two categories. The first consists of relative simple methods using elastic solutions, which is usually restricted to uniform soils. The second is to use more sophisticated numerical methods, such as the finite element method and the boundary element method. This type of methods is more versatile, but the computation effort required is much more pronounced. In this paper a method, which is relatively efficient and can be easily applied to layered soil, is proposed. In this method the raft is modeled by Mindlin plate elements. The soil is modeled by finite and infinite prism elements proposed by Wang and Gu [3], and the piles are modeled by axially loaded line elements. A computer program is developed implementing the method. Some numerical examples are presented to show the efficiency and acceptability of the method.

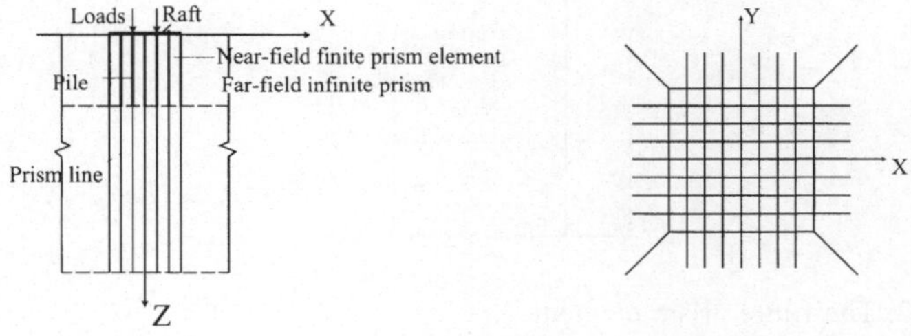

Figure 1. The model for piled raft

The numerical model

Consider a piled raft under vertical loads on layered soil Figure 1. The layering of the soil can be somewhat irregular beneath the raft (the near field), whereas horizontal layering is assumed for the soil outside the raft area (the far field). The raft is discretized into plate elements, and correspondingly the soil beneath the raft, down to an estimated compressible depth, is discretized into finite prisms. The cross section of any prism is the same as the raft element above it. In the vertical direction the prisms are divided by the soil layer interfaces. Outside the raft area the soil is discretized into infinite prism elements, which extend horizontally to infinity with their vertical dimension being the same as those beneath the raft edge. In the above discretization, attention should be paid to locate the piles along the edges of the prisms. The system is assumed to be elastic, i.e. the material behaves elastic and the slippage between the pile and the soil is neglected. To further reduce the computation cost, horizontal displacements are assumed to be negligible. The latter assumption seems reasonable for such a system under vertical loads.

Plate bending elements for the raft

In order to model both thin and moderately thick raft, plate finite elements of Mindlin type are used, in particular a 3-node triangle element and a 4-node quadrilateral element. For these lower order plate bending elements attention has to be paid to avoid shear locking. In this work, for the 4-node element we use a mixed interpolation scheme [4], whereas for the 3-node triangle plate element we use the concept of deformation decomposition [5]. The 3-node element is usually just employed along irregular raft edges, as it is, according to our experience, not as accurate as the 4-node element.

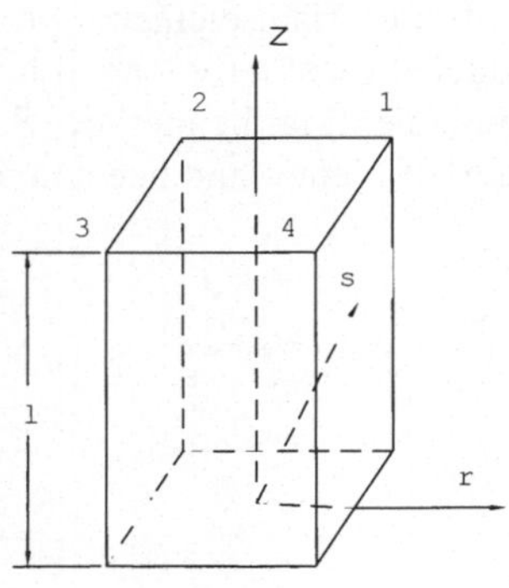

Figure 2. The finite prism element

Finite prism element for soil in the near field

A typical finite prism element is schematically shown in Figure 2, where the 4 vertical edges are numbered from 1 to 4. Along a vertical edge *i,* the vertical displacement is assumed to be

$$w_i = w_i^0\left(1-\frac{z}{l}\right) + w_i^l\left(\frac{z}{l}\right) + \sum_{k=1}^{m} A_{ik} \sin\frac{k\pi z}{l} \qquad (1)$$

where w_i^0 and w_i^l are respectively the displacements at the upper and lower end of edge i. The coefficients of the Fourier series, A_{ik}, can be conceived

as generalized displacements. Numerical comparison has shown that it is usually sufficient to take just the first two terms from the series to obtain results with acceptable accuracy [6].
With the displacements along the edges being written as equation (1), the displacements at any point inside the prism can be interpolated using the displacements at the 4 edge points lying at the same depth. The interpolation scheme is the same as for a quadrilateral element. Thus, the displacement field within a prism is completely defined in terms of w_i^0 , w_i^1 and A_{ik}. Then strains and stresse scan be discribed. By using the virtual work principle the stiffness matrix of the finite prism element can be easily derived.

Line element for piles

The pile is modeled by axially loaded line elements located along the edges of the finite prism element. The interpolation scheme is the same as that given by equation (1). The stiffness matrix can then be obtained in the same way as for common types of finite elements. In order to consider the pile tip resistance, especially for relatively short pile, we suggest that soil pile elements, corresponding to the soil underneath the pile tip, are added from the pile tip to the mesh bottom.

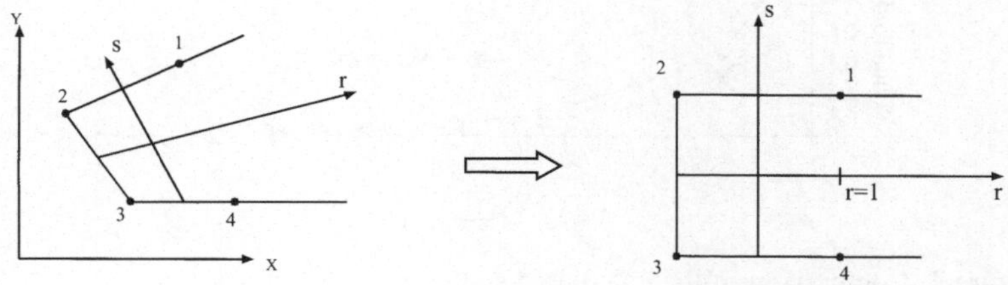

Figure 3. Plan view of the infinite prism element

Infinite prism element for soil in the far field

Figure 3 shows the plan view of an infinite prism element in the global coordinate and standardized local coordinate system. A striking difference of an infinite prism element from the finite prism element is that it extends to infinity in one or two directions. The displacement interpolation function for an infinite element must include a decaying function along the infinite direction beside the ordinary interpolation function. For the element shown in Figure 3, the interpolated displacement is written as

$$w = M_2 w_2 + M_3 w_3 \tag{2}$$

$$M_i = 0.5(1 + s_i s) f(r) \quad \text{and} \quad f(r) = 1/(2 + r) \tag{3}$$

where w_i (i=2,3) are displacements along the vertical edges, defined in a similar way as expressed in equation (1). M_i is the interpolation function for determining the displacement at any point from those at the edges. It consists of an ordinary interpolation function multiplied by the decaying function *f(r)*. This expression of f(r) has been established through comparative study. Again the same procedure can be followed to get the stiffness matrix of the infinite prism element. After the determination of the stiffness of the various components, the global stiffness matrix of the entire piled raft system can be assembled. Obviously, the global matrix retains the desired characteristic of relatively small bandwidth. This is one of the advantages of this method over those methods using elastic Mindlin solutions.

Validation and application

A computer program called TUF2 has been made implementing the above model, and a number of calculations have been performed to test the model as well as the computer program.

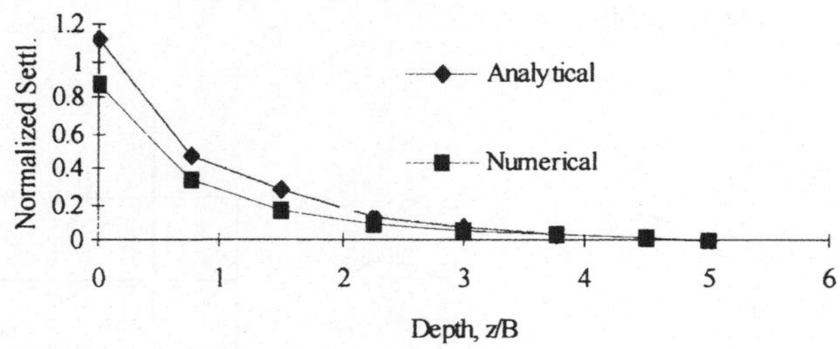

Figure 4. Comparison of the calculated settlements

In order to investigate the performance of the finite and infinite prism elements, the settlements of the elastic half space under a uniformly distributed load of $100 kN/m^2$ over a $4 \times 4 m^2$ area are calculated. The settlements at the central line calculated by TUF2 are compared with the Boussinesq solution in Figure 4, where we see that the accuracy is acceptable although it is not very accurate. It can be understood that the settlement calculated by TUF2 is always smaller than the Boussinesq solution, and the difference can be up to 25%. From engineering experience we know that the elastic theory usually exaggerates the influence zone of a load and gives larger settlements. Hence, it can be expected that the comparison of the results by TUF2 with field measurements can be better.

The model has been also applied for analyzing engineering projects. As an example, the analysis results for the piled raft foundation of a building to be constructed in Shanghai are presented. The southern sector of this building is composed of two 30-storey towers standing opposite to each other in the west-east direction with 4-storey building blocks in between. The base area is about 80×40m^2.The plan is shown in Figure 5, where piles are installed beneath the two towers. The blue areas on the left and right correspond to the core tube positions. The left and right sides of the foundation are not exactly symmetric as the raft extends by 4m out from the left most row of columns. The embedment depth is 12m. The pile is 21m long and the cross-section 0.5m×0.5m. The raft is a reinforced concrete plate with a thickness of 2m.

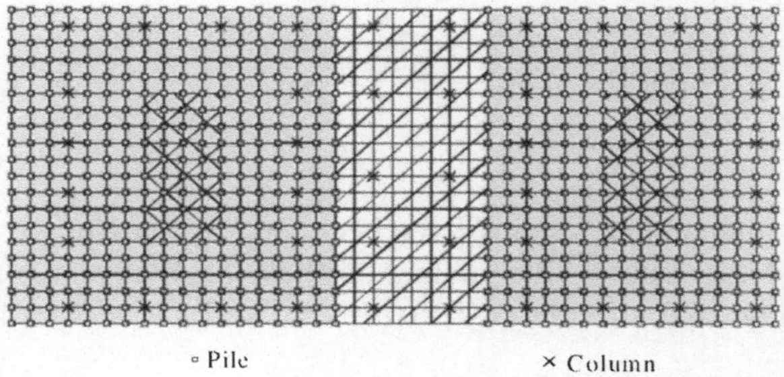

Figure 5. The mesh and the layout of the piles

Calculations are performed with the proposed method. Elements in the area of the core tubes and along the exterior walls are assumed to be rigid accounting for the stiffness of the reinforced concrete walls on the raft. The calculated settlements are shown in Figure 6. Below the left core tube the settlement is 8.1cm, and below the right core tube 8.6cm. In the middle it is 7.6cm. On the left edge the minimum settlement is 3.5cm and on the right edge 5cm. The 4m extension of the raft to the left, gives larger differential settlement. Hence, we suggested that few piles should be installed there. It coincides with results obtained by using a different approach [7].

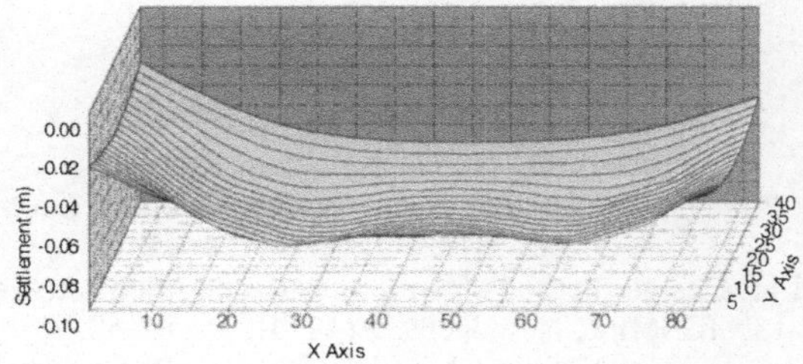

Figure 6. The settlement of the raft

The inner forces of the raft have been calculated. Figure 7 shows the distribution of the bending moment M_y. As the bending moments in the core tube area and along the raft edges are also carried by the walls, they are not shown. The largest bending moments in the raft, occurring near the edge of the core tube, are about 8000kNm/m for M_y and 4500kNm/m for M_x. And the largest negative bending moment is about –4000kNm/m, occurring in the middle of the raft.

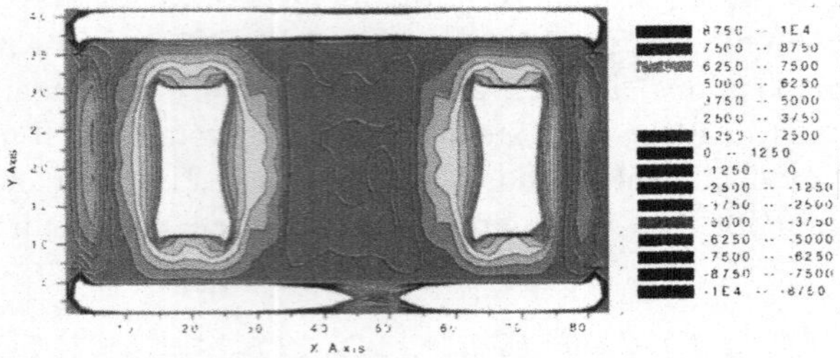

Figure 7. The distribution of M_x

Extensive comparative studies have been carried out to investigate some influence factors, such as the number of piles, the pile installation area etc. on the average and differential settlement as well as the inner forces of the raft.

Concluding remarks

A model for the analysis of piled raft foundation has been briefly reported. Salient features of the model are the assumption of negligible horizontal displacements and the use of finite and infinite prisms for modeling the soil. In this model the banded nature of the global matrix is retained, and layered soil deposits can be easily considered. Numerical calculations have shown that it is relative efficient and can give results with acceptable accuracy. As further developments the influence of the rigidity of the superstructure and the non-linearity of the system, especially at the pile-soil interface, are recommanded.

References

1. Poulos, H.G.: Piled raft foundations: design and applications, *Geotechnique* 51(2) (2001), pp. 95-113
2. Randolph, M.F.: Design methods for pile group and piled rafts: state of the art report, Proc. 13th ICSMFE, New Delhi 5 (1994), pp. 61-68.
3. Wang, W. and Gu, X.L.: Numerical models for pile-raft foundation analysis, *J. of Civil Eng.*, Vol. 31 (1998), pp. 287-299.
4. Bathe, K.J. and Dvorkin E.N.: A four-node plate element based on Mindlin/ Reissner plate theory and a mixed interpolation, *J. Num. Meth. Eng.*, Vol. 21 (1985), pp. 367-383
5. Belytschko, T., Stolarski H. and Carpenter N.: A C^0 triangular plate element with one-point quadrature, *J. Num. Meth. Eng*, Vol. 20 (1984), p: 787-802
6. Gao, H.Z.: *Numerical analysis of soil-piled raft interaction*, M-Sc thesis, Tsinghua University, June 1999.
7. Song, E.X. and Gao H.Z.: A practical method for deformation analysis of pile-raft foundation, Proc. 5th Inter. Conf. *Tall Buildings*, Hong Kong (1998), pp.966-971.

On the Stability of Unlined Tunnels

PIETER A. VERMEER, URS VOGLER
Institute of Geotechnical Engineering, University of Stuttgart, Germany

This paper addresses first of all the stability of a fully unlined circular tunnel. Attention is focused on a homogenous Mohr-Coulomb material under drained conditions. The unlined tunnel is assumed to be supported by an internal pressure and a formula for the collapse pressure is derived. The soil friction angle is shown to be extremely important, as otherwise there is very little arching. For highly frictional material, an arching ground ring is shown to develop around the tunnel. This confirms common ideas on tunnelling, but the novel finding is the extreme influence of the friction angle as well as the role of the cohesive strength. Comparison is made with data for lined tunnels as published in a separate paper. Finally attention is focused on the accuracy of FEM for limit loads. To this end upper and lower bound solutions by other authors are considered. These data testify directly the accuracy of present solutions.

Introduction

Several authors, e.g. Vermeer & Van Langen [3], have shown that the nonlinear elastic-plastic finite-element method is well-suited to predict collapse loads of geotechnical structures. For this purpose, pre-failure deformations are not of great importance and assumed to be linear elastic. In addition to the elasticity modulus E and the Poisson´s ratio ν, there are three material parameters for the plastic behaviour: cohesion c´, angle of friction φ´ and angle of dilatancy ψ. This constitutive model will be used to compute the minimal supporting pressure of tunnels.

Considering a perfectly plastic constitutive model, failure is defined as continuing deformation without any (further) change of stress. For models with an associated flow rule $(\psi = \varphi')$, the corresponding collapse load is unique, i.e. independent of elasticity parameters and the initial stress-state. On using a non-associated flow rule $(\psi \neq \varphi')$ as done in the present study there is no general proof of uniqueness and one will have to investigate the particular problem considered. For this reason, various sensitivity analyses have been carried out for tunnel stability problems and results showed no influence of K_0, ψ or elasticity parameters on the collapse load.

In practice unlined tunnels are not as frequent as lined tunnels, but they do occur and are thus of practical interest. Besides they are of theoretical interest, as they involve a state of plane strain. As a consequence, the unlined tunnel constitutes a 2D-problem which can be solved both analytically (Atkinson & Potts [1]) and semi-analytically (Lyamin & Sloan [7]). Such solutions are of importance for the validation of numerical FE-simulations as presented in the present paper.

Arching around unlined Tunnels

As the friction angle appeared to have a tremendous effect in case of lined tunnels (Vermeer & Ruse [4]), its influence will also be investigated for unlined tunnels. To this end a tunnel with a relative cover of $C/D = 4$ and a relative cohesion of $c'/\gamma D = 0.125$ will be considered, where D is the tunnel diameter and γ the specified soil weight. Attention will be focused on soils with three different friction angles of 0°, 20° and 40° respectively. The soil dilatancy angle is taken equal to zero and initial stresses are geostatic with $K_0 = 1$. As for the lined tunnels, the supporting tunnel pressure is initially in equilibrium with these geostatic stresses and then stepwise decreased to compute the failure pressure p_f. In this manner ground response curves (Pacher [9]) are obtained. Ground response curves are well-known from the literature on the New Austrian Tunnelling Method (e.g. Rabcewicz et al [2]). Rather then on the entire ground-response-curve, the present study focuses on collapse pressures p_f and associated failure mechanisms.

Computational results show a tremendous influence of the friction angle. This is well observed from the stress fields as depicted in Figure 1. For frictionless materials, a rotation of principal stresses is observed (Fig. 1a), but there is no arching and a high failure stress of $p_f/\gamma D = 4.03$ is obtained, i.e. only slightly below the initial value of $p_0/\gamma D = 4.5$. In contrast, one finds a very low value of $p_f/\gamma D = 0.03$ for the highly frictional soil with $\varphi' = 40°$. It appears from Figure 1c that this is due to considerable arching. The top of the arching ground ring appears to be well above the tunnel and in between

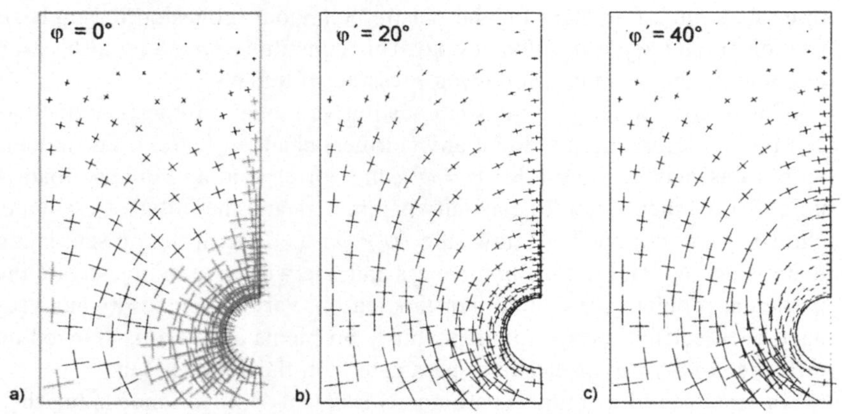

Figure 1. Computed principal stresses at failure for $C/D = 4$ and $c'/\gamma D = 0.125$

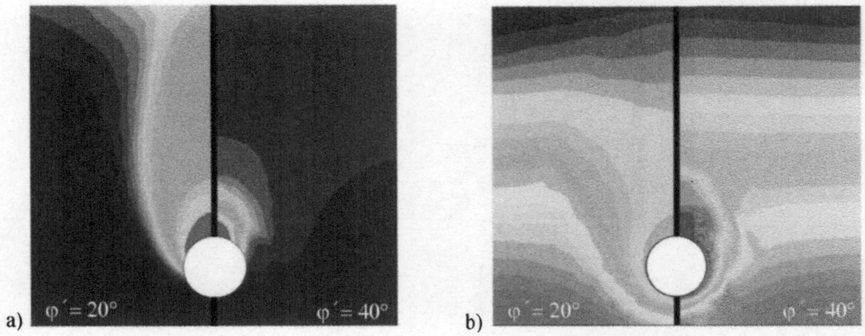

Figure 2. Velocity contours (a) and mean stress (b) at failure for $C/D = 4$

stresses have nearly vanished. Hence the circular ground opening is observed to be more or less protected by an eccentric elliptical pressure ring. This confirms common ideas on tunnelling, but the novel finding is the considerable influence of the friction angle. Indeed, for a low friction angle of 20° the arching ground ring is not very pronounced and difficult to observe, as can be seen in Figure 1b.

A further comparison between data for $\varphi' = 20°$ and $\varphi' = 40°$ is presented in Figure 2 by showing contour lines of displacement increments and mean principal stresses $p = (\sigma_1 + \sigma_2 + \sigma_3)/3$. Figure 2a shows that the impact of arching is also reflected in the collapse mechanisms. For a low friction angle of 20°, the failure pattern extends to the surface and a chimney-like collapse mechanism is observed. For highly frictional ground, however, material movements do not extend to the ground surface at all. Instead, one observes local ground movements inside the arching pressure ring. For high friction angles, the contour lines in Figure 2b are eccentric ovals around the tunnel and one observes the arching ground ring most directly. On the other hand the low frictional material does not lead to closed stress contour lines.

Stability Numbers for unlined Tunnels

Considering similarities between the face stability of a lined tunnel (Vermeer & Ruse [4]) and the stability of an unlined tunnel, it would seem the same equation applies to both type of tunnel. It may obviously not be expected that stability numbers for a lined tunnel would also apply to an unlined tunnel, but it is reasonable to assume an equation of the form

$$p_f = \gamma D N_\gamma - c' N_c + q N_q \qquad \text{with } N_c = cot\varphi' \text{ and } N_q = 0$$

where p_f is the failure pressure, q is a possible surface charge and N_γ, N_c and N_q are friction dependent stability numbers. In order to validate this idea, attention will be focused first of all on the simple problem of a circular tunnel in dry cohesionless soil ($c'=0$) and no surcharge on the ground surface ($q=0$).

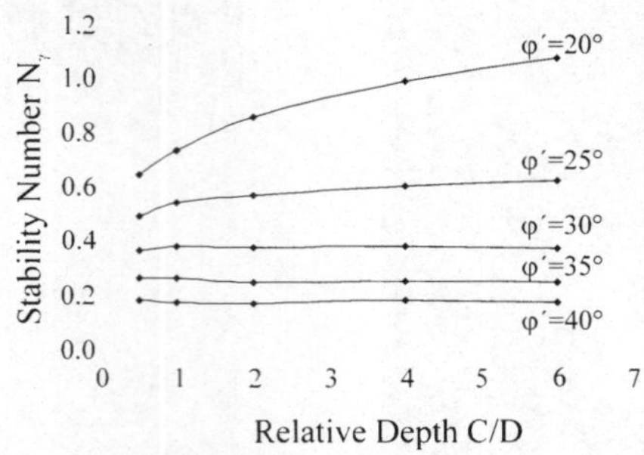

Figure 3. Computed N_γ-values for unlined tunnels

In such a case the above equation reduces to $p_f = \gamma D N_\gamma$ and it is relatively easy to assess the stability number N_γ. To this end a series of numerical simulations with various different friction angles and various different tunnel depths have been performed. Computational results, as shown in Figure 3, show that the relative depth *C/D* is of little importance, at least for friction angles larger than 25°.

The data of Figure 3 yield the N_γ-curves of Figure 4, which converge to a unique single curve for a friction angle of 30°. For $\varphi' = 25°$ differences are already small and this angle might well be regarded as the threshold value for depth-independent N_γ-values. These FE-curves are in an agreement with the theoretical bound solution by Atkinson & Potts [1], in the sense that they are indeed above the bounding line.

In Figure 5 all computational data are repeated and indicated by dots. For unlined tunnels N_γ-values appear to be at least twice as large as the ones for lined tunnels (Vermeer & Ruse [4]). The full curves in Figure 5 represent the analytical approximations:

$N_\gamma = 0.6 \cot^2 2\varphi' + 0.18$ for unlined tunnels

$N_\gamma = 1/9 \cot \varphi' - 0.05$ for lined tunnels

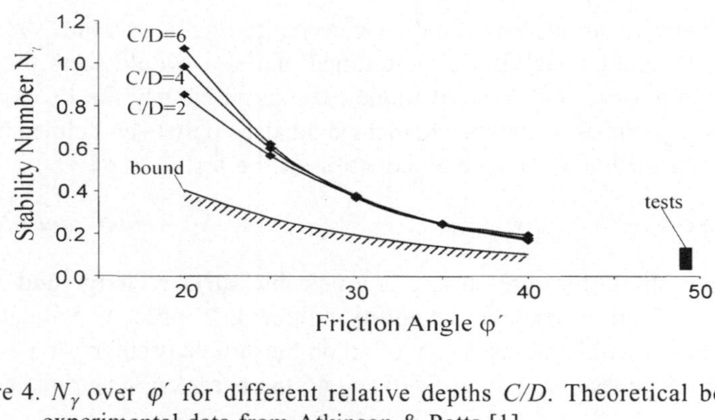

Figure 4. N_γ over φ' for different relative depths *C/D*. Theoretical bound and experimental data from Atkinson & Potts [1]

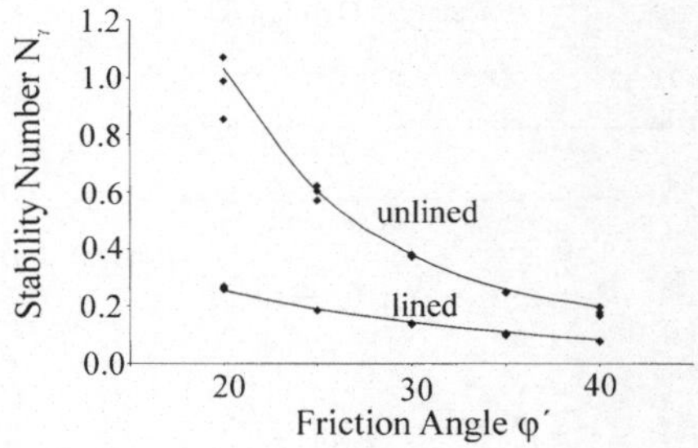

Figure 5. Comparison of values for lined and unlined tunnels

It can be concluded that there is a local mode of failure around the tunnel roof for $\varphi' \geq 35°$ and $C/D \geq 3$. For such cases, one observes strong arching and the stability number N_γ is independent of depth. Other additional FE-analyses have been carried out to check the hypotheses $N_c = \cot\varphi'$ and $N_q = 0$. It appears that these simple expressions may also be used for high and moderate friction angles, but certainly not for small angles below 25°.

Accuracy of the FE-Method

In order to appreciate data from FE-analyses, one has to verify and validate the accuracy of this method. A theoretical verification is possible by checking whether or not computed collapse loads lie between lower and upper bound solutions. Upper and lower bound theorems apply to perfectly plastic materials whose flow rules are associated, i.e. $\psi = \varphi'$, where ψ is the angle of dilatancy. In general the flow rule for real geomaterials is non-associated with $\psi < \varphi'$. Nevertheless it can be shown that an upper bound calculated assuming $\psi = \varphi'$ is also an upper bound for the case where $\psi < \varphi'$. Hence, bound solutions as presented by Atkinson & Potts [1], as shown in Figure 4, apply and indicate

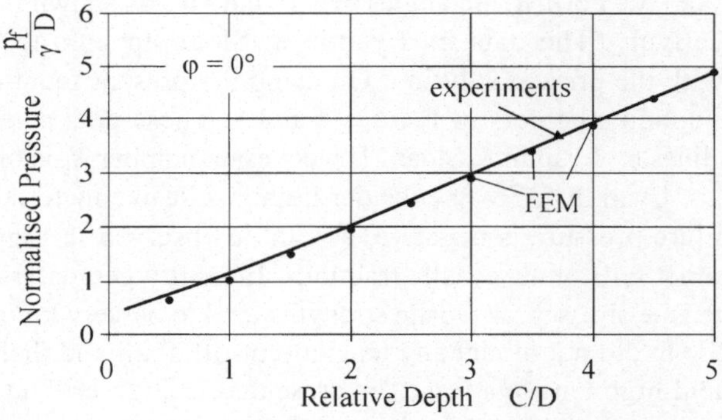

Figure 6. Normalised failure pressures for a lined tunnel. The FE results are shown as dots. Mair´s [6] experimental results are plotted as a line.

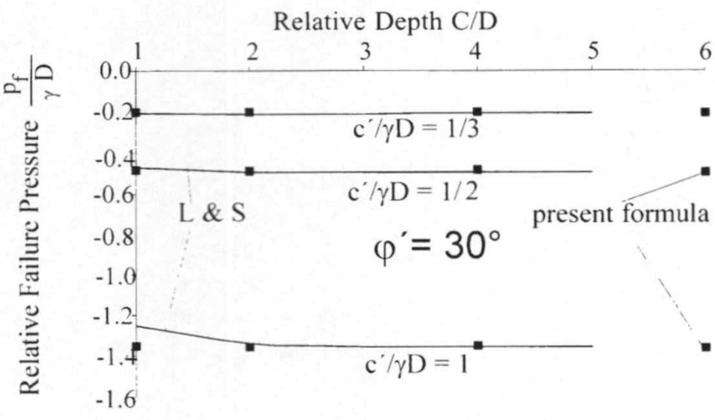

Figure 7. Comparison of present formula with solutions by Lyamin & Sloan [7] for unlined tunnels

that the FE-data is accurate. In addition experimental data appear to be in good agreement with present FE-results. This is first of all observed in Figure 4, where experimental data is indicated by a short vertical bar. Moreover Mair [6] presented results of centrifugal modelling for lined tunnels in undrained clay, which can be used for validation. Mair's results are shown by the curve in Figure 6, whilst FE-data are indicated by dots. The striking fit of both sets of data demonstrates the accuracy of 3D FE-analyses as used for the face stability of lined tunnels.

The accuracy of 2D FE-analyses, as used for unlined tunnels, can be judged on the basis of upper and lower bound solutions. Highly accurate solutions that bound the true collapse load solution to within a few per cent were presented by Lyamin & Sloan [7]. In fact, his upper and lower bounds are so close, that they virtually represent the true solution. For this reason one may present their solutions by single curves as done in Figure 7 and 8, where upper and lower bounds are simply averaged. The paper by Lyamin & Sloan involves a parametric study with friction angles of 10°, 20°, 30° and 40° respectively. For a validation of the present study, low friction angles cannot be considered, as here we have none or little data on the stability numbers N_c and N_γ. For friction angles of 30° and 40°, as shown in Figures 7 and 8 respectively. The data by Lyamin & Sloan appears to be in full agreement with the present solution. On using the present functions for N_γ and N_c, one obtains the dots in Figures 7 and 8. These dots coincide fully with the full lines by Lyamin & Sloan. The cohesion number N_c is of particular importance, as Lyamin & Sloan consider highly cohesive materials. In such cases the failure pressure is negative, as can be observed in Figures 7 and 8, i.e. the tunnel fails under tensile tractions. In reality geomaterials cannot stand large tensile stresses, as tensile strengths tend to be very low. However, Lyamin & Sloan did not model any tension-cut-off at all and their analyses may thus yield high tensile stresses as presented in Figures 7 and 8. It can be seen from these figures that the dots (of the present analysis) are virtually exactly on the lines by Lyamin & Sloan.

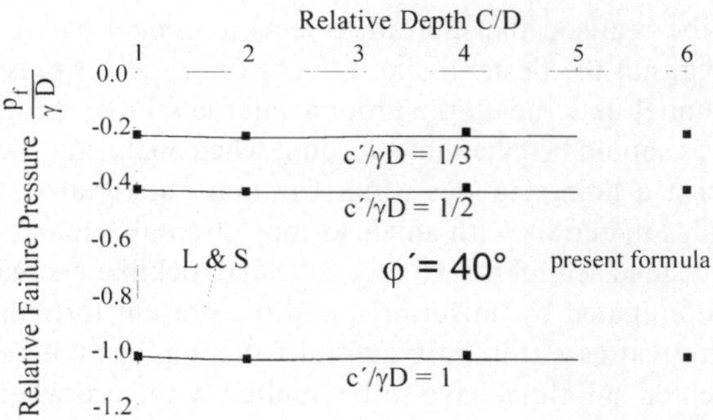

Figure 8. Comparison of present formula with solutions by Lyamin & Sloan [7] for unlined tunnels

Conclusions

FE-simulations show that tunnel stability in frictional material is dominated by arching. As a consequence, the effective supporting pressure required to prevent collapse of a tunnel is relatively small. For friction angles beyond a particular threshold value, the arching is so significant that the failure pressure is independent of tunnel depth and can be computed as $p_f = \gamma D N_\gamma - c' N_c$, where N_γ and N_c are functions of the friction angle alone.

Both for lined and unlined tunnels, the cohesion number is $N_c = cot\varphi'$ for $\varphi' > \varphi_t'$, where φ_t' represents threshold values. For lined circular tunnels, the three-dimensional arching around the tunnel face is so strong that a relatively low threshold value was detected, i.e. a threshold friction of $\varphi_t' = 20°$ (Vermeer & Ruse [4]). For unlined circular tunnels, two dimensional arching around the tunnel roof was observed. Two-dimensional arching is obviously weaker than three-dimensional arching and this is reflected in the threshold value $\varphi_t' = 25°$. It is convenient to have $N_c = cot\varphi'$ for both types of tunnels, as it implies that this simple expression can also be used for partially lined tunnels. Such situations are common for NATM-tunnelling and the associated problem of tunnel-heading stability is treated in a separated paper (Vermeer & Ruse [4]).

For highly cohesive materials, the term $-c'N_c$ may lead to highly negative pressures. In reality, geomaterials have a particular tensile strength and the tensile strength criterion implies a lower bound of $p_f \geq -TS$, where TS is the tensile strength of the material considerer. Such a situation also implies failure modes that differ from the ones considered in the present paper. The p_f-values as computed are thus associated with shear failure modes and highly negative values imply that a supporting pressure can be omitted. In competent hard rock this will definitely be the case and tunnels can be constructed without any support. The possibility of open face tunnelling will even present itself for many hard soils, as three-dimensional arching renders low values for N_γ. In such cases the present theory can be extended and used for the computation of safety factors (Vermeer & Ruse [4]).

It should be realised that not all aspects of tunnel stability have been addressed, e.g. not the destabilising effect of pore water pressures. Indeed an unlined tunnel underneath the groundwater label may act as a drain and seepage forces should be taken into account when analysing stability. Finally is recalled that a homogeneous Mohr-Coulomb material was considered, which excludes materials with an anisotropy strength such as jointed rocks and heavily bedded sediments. Highly cohesive rocklike geomaterials would seem to be dominated by anisotropy, and the present formula should than be used with great care. Similarly, strain-softening has not been modelled, so that presented solutions have to be applied with care when considering practical tunnel projects.

References

1. Atkinson, J.H., Potts, D.M.: Stability of a shallow circular tunnel in cohesionless soil. *Géotechnique* 27 (1977), No. 2, pp.203-215.
2. Rabcewicz, L., Gosler, J., Hackl, E.: Die Bedeutung der Messung im Hohlraumbau, Teil I (in German). *Der Bauingenieur* 47 (1972), No. 7, pp.225-237.
3. Vermeer, P.A., van Langen, H.: Soil collapse computations with finite elements. *Ingenieur-Archiv* 59 (1989), pp.221-236.
4. Vermeer, P.A., Ruse, N.: Tunnel Heading stability in Homogeneous Ground. *Géotechnique* (presented for publication 2002).
5. Verruijt, A.: Deformations of an elastic half plane with circular cavity. Int. J. Solids Struct. 21 (1998), pp.2795-2804.
6. Mair, R.J.: *Centrifugal modelling of tunnelling construction in soft clay.* PhD thesis, Universtity of Cambridge, UK, 1979.
7. Lyamin, A.V., Sloan, S.W.: Stability of a plane strain circular tunnel in a cohesive-frictional soil. *Developments in theoretical Geomechanics*. A.A. Balkema Publishers, Rotterdam, 2000, pp.139-153.
8. Verruijt, A., Booker, J.R.: Complex variable analysis of Mindlin´s tunnel problem. *Developments in theoretical Geomechanics*. A.A. Balkema Publishers, Rotterdam, 2000, pp.3-22.
9. Pacher, F.: Deformationsmessungen im Versuchsstollen als Mittel zur Erforschung des Gebirgsverhaltens und zur Bemessung des Ausbaus (in German). *Felsmech. und Ing.-Geol. Suppl.* I, 1964.

GEODYNAMICS

Moving Line Load on an Elastic Medium with Horizontal Constraints

CARLOS J. CORNEJO CÓRDOVA
TUDelft, Delft, the Netherlands

Analytical solutions of transient wave problems can be obtained by using integral transformation methods, especially Laplace and Fourier transformations. The mathematical problem is then reduced to the determination of the inverse transformation. For this purpose the Cagniard-De Hoop method [1] is applied. The propagation of transient waves in a half-space due to a moving load on the surface has been investigated by [2], [3] and [4].

In problems consisting of an elastic half-space in which loads of vertical direction are applied on the surface only, it can be assumed that the vertical displacements will be notably larger than the horizontal displacements. Then an approximate method of solution can be developed. In this method the horizontal displacements are disregarded, so that the only actual displacement is the vertical displacement. In order to determine the solution of problems under these assumptions the medium is referred to as a material with horizontal constraints. The solution of the Boussinesq problem was credited to [5] for materials satisfying these assumptions. The solution of a vertical line load of finite length acting at the surface was given by [6]. Transient wave problems were examined by [7] and [8].

The purpose of this paper is to obtain a closed-form solution of the vertical displacement for a vertical moving line load acting on the surface of a half-space, depicted in Figure 1. The inverse Fourier and Laplace transformations can be found in a simple way if the Cagniard-De Hoop method is used.

In a rectangular coordinate system, consider the half-space $z > 0$ and $-\infty < x < \infty$, with the bounding surface $z = 0$. For a two-dimensional problem the governing scalar wave equation of motion in the vertical direction of a material with horizontal constraints, the so called Westergaard's equation of motion, is

$$\mu\partial_{xx}^2 u_z + (\lambda + 2\mu)\partial_{zz}^2 u_z = \rho\partial_{tt}^2 u_z, \tag{1}$$

where μ and λ are Lamé's constants, ρ is the density of the material and $u_z = u_z(x, z, t)$ is the vertical displacement. This can be rewritten as

$$c_s^2\partial_{xx}^2 u_z + c_p^2\partial_{zz}^2 u_z = \partial_{tt}^2 u_z, \tag{2}$$

where $c_p = ((\lambda + 2\mu)/\rho)^{1/2}$ is the velocity of the dilatation waves and $c_s = (\mu/\rho)^{1/2}$ is the velocity of the distortional waves. The corresponding vertical stress is $\sigma_{zz} = (\lambda + 2\mu)\partial_z u_z$.

Moving line load

Wave motion in a half-space due to a surface moving line load at constant speed v is shown in Figure 1. The boundary condition on the surface $z = 0$ is

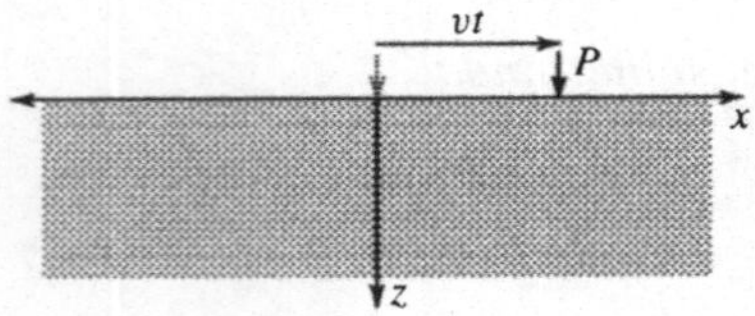

Figure 1. Moving line load.

represented by

$$z = 0 \; : \; \sigma_{zz} = -P\delta(x - vt)H(t); \;\; v > 0, \tag{3}$$

where P is a line force per unit of length, $\delta(x - vt)$ is the Dirac function and $H(t)$ denotes the Heaviside unit step function. For $t = 0$, it is assumed that the displacements and velocities are zero, i.e. $u_z = 0$ and $\partial_t u_z = 0$.

The Laplace transformation with respect to t is defined as,

$$\hat{u}_z(x, z, s) = \int_{t=0}^{\infty} u_z(x, z, t) \exp(-st)\, dt, \tag{4}$$

where s is a real and positive parameter. From (2), one obtains

$$c_s^2 \partial_{xx}^2 \hat{u}_z + c_p^2 \partial_{zz}^2 \hat{u}_z = s^2 \hat{u}_z. \tag{5}$$

With the aid of the property $\delta(x - vt) = v^{-1}\delta(t - xv^{-1})$, and the Laplace transformation

$$\int_{t=0}^{\infty} \exp(-st)\delta(t - xv^{-1})\, dt = \exp(-sxv^{-1})H(x), \tag{6}$$

the boundary condition (3) in the Laplace domain is

$$z = 0 \; : \; \hat{\sigma}_{zz} = -Pv^{-1} \exp(-sxv^{-1})H(x). \tag{7}$$

The Fourier transformation of $\hat{u}(x, z, s)$ with respect to x, according to

$$\tilde{u}_z(\alpha, z, s) = \int_{x=-\infty}^{\infty} \hat{u}_z(x, z, s) \exp(is\alpha x)\, dx, \tag{8}$$

is introduced. The factor s in the argument of the exponential function has been introduced for convenience. Substituting (8) in (5) and (7), gives

$$\partial_{zz}^2 \tilde{u}_z - s^2 \beta^2 \tilde{u}_z = 0, \tag{9}$$

where $\beta = (\alpha^2 + c_s^{-2})^{1/2} c_s / c_p > 0$, with the boundary condition

$$z = 0 \; : \; \tilde{\sigma}_{zz} = P \frac{v^{-1}}{s(i\alpha - v^{-1})}. \tag{10}$$

The solution of (9) that remains bounded for $z \to \infty$ is

$$\tilde{u}_z(\alpha, z, s) = \tilde{A}\exp(-s\beta z), \tag{11}$$

where $\tilde{A}$ is a parameter. From (10) and (11), it follows that

$$\tilde{A} = -\frac{P}{s(\lambda + 2\mu)\beta}\frac{v^{-1}}{(i\alpha - v^{-1})}. \tag{12}$$

Using the inverse of the Fourier transformation the solution of the problem is, expressed in the Laplace transformation,

$$\hat{u}_z(x, z, s) = -\frac{P}{s(\lambda + 2\mu)}\frac{1}{2\pi}\int_{\alpha=-\infty}^{\infty}\frac{\exp(-is\alpha x - s\beta z)v^{-1}}{\beta(i\alpha - v^{-1})}\,d\alpha. \tag{13}$$

The parameter of integration with respect to α is replaced by the parameter of integration with respect to p, defined by $p = i\alpha$. It follows that

$$\hat{u}_z(x, z, s) = -\frac{P}{s(\lambda + 2\mu)}\frac{1}{2\pi i}\int_{p=-i\infty}^{i\infty}\frac{\exp(-s(px + \beta z))v^{-1}}{\beta(p - v^{-1})}\,dp, \tag{14}$$

with $\beta = (c_s^{-2} - p^2)^{1/2}c_s/c_p$ and $\Re[\beta] > 0$. The integration with respect to p along the imaginary p-axis is replaced with one along the curved path, defined by $\chi = px + \beta z$, where χ is considered to be a real and positive parameter. The new path is assumed to consist of the two hyperbolic arcs p_1 and p_2, with

$$p_{1,2} = x\chi/r_n^2 \pm inz(\chi^2 - r_n^2/c_s^2)^{1/2}/r_n^2; \quad r_n/c_s \le \chi < \infty. \tag{15}$$

where $n = c_s/c_p$ and $r_n = (x^2 + n^2z^2)^{1/2}$. Substitution of (15) into $\chi = px + \beta z$ leads to the following values of β along the two arcs

$$\beta_{1,2} = n^2z\chi/r_n^2 \mp inx(\chi^2 - r_n^2/c_s^2)^{1/2}/r_n^2. \tag{16}$$

At this stage, the analysis of the complex p-plane is split into two cases: one, the velocity of the load is smaller than the velocity of S-waves ($v < c_s$) and two, the velocity of load is larger than the velocity of S-waves ($v > c_s$).

For $v < c_s$:

In the subsonic case ($v < c_s$) the new path of integration is depicted in Figure 2. The contributions of the arcs connecting the imaginary axis and the new path vanish. Therefore the integral along the imaginary p-axis, $\Re[p] = 0$ is equivalent to the integral along the hyperbolic arcs $p = p_1$ and $p = p_2$. The expression (13) is now transformed into

$$\hat{u}_z(x, z, s) = -\frac{P}{s\pi(\lambda + 2\mu)}\int_{\chi=r_n/c_s}^{\infty}\Im\left[\frac{v^{-1}}{(p_\perp - v^{-1})}\frac{1}{\beta}\partial_\chi p_\perp\right] \times \exp(-s\chi)\,d\chi; \quad v < c_s. \tag{17}$$

This expression is carried out for the case $v < c_s$ where the simple pole $p = v^{-1}$ is located to the right of the branch point $p = c_s^{-1}$. Because the region in the complex p-plane in between the imaginary p-axis and the new path is free from

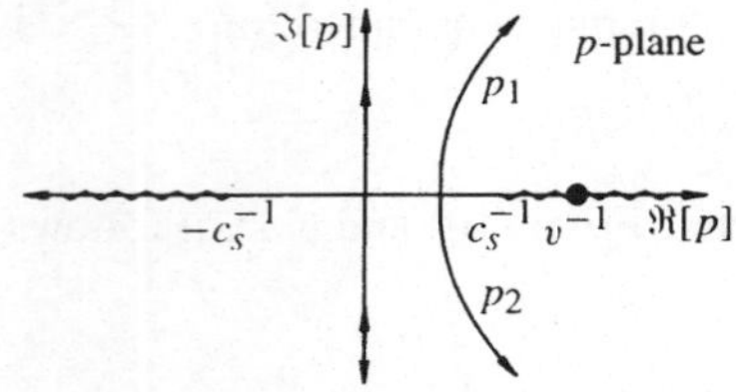

Figure 2. Path of integration in the complex p-plane: subsonic case ($v < c_s$).

singularities, then the integrated of equation (17) is not singular. In addition, branch cuts can be introduced along the real p-axis, from $-\infty$ to $-1/c_s$ and $1/c_s$ to ∞, see Figure 2. With the aid of the following expressions

$$\partial_\chi p = \pm i\beta/(n(\chi^2 - r_n^2/c_s^2)^{1/2}), \tag{18}$$

$$\frac{v^{-1}}{(p - v^{-1})}\frac{1}{\beta}\partial_\chi p = \frac{(nz(\chi^2 - r_n^2/c_s^2)^{1/2} \pm i(x\chi - v^{-1}r_n^2))}{((x\chi - v^{-1}r_n^2)^2 + n^2z^2(\chi^2 - r_n^2/c_s^2))} \times \frac{v^{-1}r_n^2}{n(\chi^2 - r_n^2/c_s^2)^{1/2}}, \tag{19}$$

in which (19) is substituted into (17), and changing the parameter of integration χ with the parameter of integration q defined as $\chi = qr_n/c_s$ and applying the shift rule for the Laplace transformation to the function $\exp(-s\chi)/s$, the time-domain expression for the vertical displacement becomes

$$u_z(x, z, t) = \frac{P}{\pi(\lambda + 2\mu)n}\int_{q=1}^{t/t_0} \frac{(xC_1 + n^2z^2)}{(C_1^2 + n^2z^2(1 - v^2c_s^{-2}))C_2}\,dq \times H(t - t_0); \quad v < c_s, \tag{20}$$

where $t_0 = r_n/c_s$ is the arrival time of the wave, $C_1 = x - r_n v c_s^{-1} q$, and

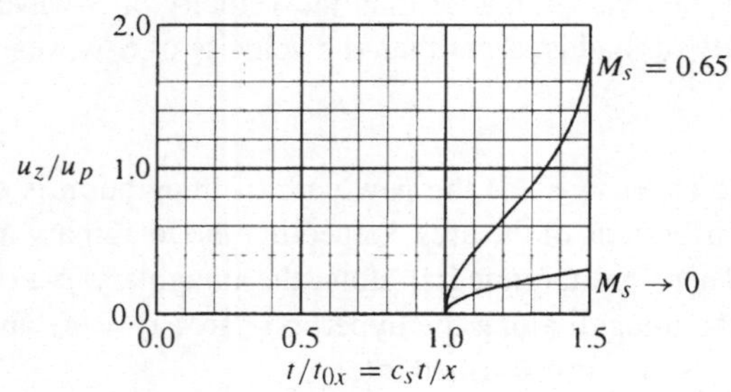

Figure 3. Dimensionless vertical displacement u_z/u_p: $z \to 0$, $vt < x < c_s t$.

$C_2 = (q^2 - 1)^{1/2}$. The vertical displacement for $z \to 0$ is

$$u_z(x, 0, t) = \frac{P}{\pi(\lambda + 2\mu)n}\frac{1}{\sqrt{1 - M_s^2}}\ln\left(\frac{1 - M_s + C_3\sqrt{1 - M_s^2}}{1 - M_s - C_3\sqrt{1 - M_s^2}}\right) \times H(t - t_{0x}); \quad vt < x < c_s t, \tag{21}$$

with $M_s = v/c_s$, $C_3 = ((t/t_{0x} - 1)(t/t_{0x} + 1))^{1/2}$, $t_{0x} = x/c_s$ and $u_p = P/((\lambda+2\mu)n)$. Figure 3 shows the dimensionless vertical displacement u_z/u_p when $z \to 0$ and $vt < x < c_s t$.

For $v > c_s$:

In the supersonic case ($v > c_s$) the analysis of the complex p-plane is split into two possible cases. In the first, the contribution from the pole $p = v^{-1}$ is not required, as it is located outside the region in between the imaginary p-axis and the new path, depicted by the left hand side of Figure 4. In the second, the process of path deformation requires a contribution from the pole $p = v^{-1}$, as in this case it is located within the region in between the imaginary p-axis and the new path, depicted by the right hand side of Figure 4. The point of intersection

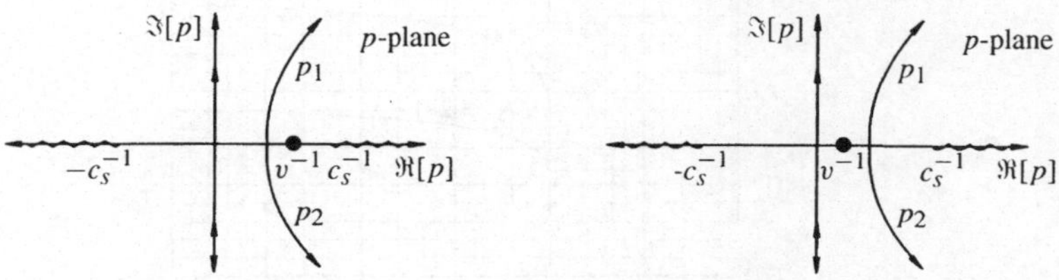

Figure 4. Path of integration in the complex p-plane: supersonic case ($v > c_s$).

of the new path p for $r_n/c_s \le \chi < \infty$ with the real p-axis is located to the right of the pole $p = v^{-1}$, that is for $x^2(v^2/c_s^2 - 1) > n^2z^2$. Applying the residue theorem to the integral on the right hand side of (14) the vertical displacement is, expressed in the Laplace transformation,

$$\begin{aligned}\hat{u}_z(x,z,s) =& \frac{P}{(\lambda+2\mu)} \frac{v^{-1}\exp(-s(xv^{-1} + n(c_s^{-2} - v^{-2})^{1/2}z))}{sn(c_s^{-2} - v^{-2})^{1/2}} \\ & - \frac{P}{s\pi(\lambda+2\mu)} \int_{\chi=r_n/c_s}^{\infty} \Im\left[\frac{v^{-1}}{(p_\perp - v^{-1})}\frac{1}{\beta}\partial_\chi p_\perp\right] \\ & \times \exp(-s\chi)d\chi; \quad v > c_s \text{ and } x > nz/(v^2/c_s^2 - 1)^{1/2}. \quad (22)\end{aligned}$$

Substituting (19) into (22), changing the parameter of integration χ with q defined as $\chi = qr_n/c_s$, and applying the shift rule for the Laplace transformation to the function $\exp(-s\chi)/s$, the vertical displacement is

$$\begin{aligned}u_z(x,z,t) =& \frac{P}{(\lambda+2\mu)n} \frac{1}{(v^2/c_s^2 - 1)^{1/2}} H(t - t_v) \\ & + \frac{P}{\pi(\lambda+2\mu)n} \int_{q=1}^{t/t_0} \frac{(xC_1 + n^2z^2)}{(C_1^2 + n^2z^2(1 - v^2c_s^{-2}))C_2} dq \\ & \times H(t - t_0); \quad v > c_s \text{ and } x > nz/(v^2/c_s^2 - 1)^{1/2}, \quad (23)\end{aligned}$$

with $t_v = x/v + n(c_s^{-2} - v^{-2})^{1/2}z > 0$. The influence of the first factor on the right hand side of (23) is restricted to its spatial support $x > nz/(v^2/c_s^2 - 1)^{1/2}$.

In this region, the wave associated with t_v arrives earlier than the one associated with $t_0 = r_n/c_s$. The vertical displacement for $z \to 0$ is

$$\begin{aligned} u_z(x,0,t) = & \frac{P}{(\lambda+2\mu)n} \frac{1}{(M_s^2-1)^{1/2}} H(t-x/v) \\ & + \frac{P}{\pi(\lambda+2\mu)n} \frac{1}{(M_s^2-1)^{1/2}} \arccos\left(\frac{M_s - t/t_{0x}}{M_s t/t_{0x} - 1}\right) \\ & \times H(t-t_{0x}); \quad v > c_s \text{ and } x > 0. \end{aligned} \tag{24}$$

Figure 5 shows the dimensionless vertical displacement u_z/u_p when $z \to 0$. The influence of the residue term can be observed at the arrival time $t = x/v = t_{0x}/M_s$. This means that a finite jump discontinuity occurs.

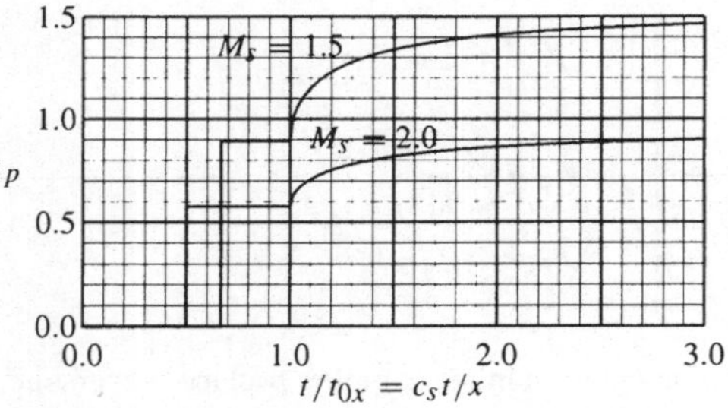

Figure 5. Dimensionless vertical displacement u_z/u_p: $z \to 0$, $v > c_s$ and $x > 0$.

References

1. De Hoop, A.T.: A Modification of Cagniarg's Method for Solving Seismic Pulse Problems. *Appl. Sci. Res.* B8m, (1960), pp. 349-356.
2. Gakenheimer, D.C. and Miklowitz, J.: Transient Excitation of an Elastic Half Space by a Point Load Traveling on the Surface. *J. Appl Mech.* 36, (1969), pp. 505-515.
3. Miklowitz, Julius: *The Theory of Elastic Waves and Waveguides.* North-Holland Publishing Company Amsterdam, 1978.
4. Bakker, Maarten C.M.: *Non-Destructive Inspection Techniques Based on Elastodynamics: Development and Validation.* Ph.D. Thesis, Delft University of Technology, Delft, The Netherlands, 2000.
5. Westergaard, H.M.: *A Problem of Elasticity Suggested by a Problem in Soil Mechanics: Soft Material Reinforced by Numerous Strong Horizontal Sheets, in Contributions to the Mechanics of Soils, Stephen Timoshenko 60th Anniversary Volume.* Macmillan, New York, 1938.
6. Fadum, Ralph E.: Influence Values for Estimating Stresses in Elastic Foundations. *Proc. Second Int. Conf. Soil. Mech. and Found. Eng.* 3, Rotterdam, 1948.
7. Barends, F.B.J.: Elasto-Dynamica Response of a Half Plane. *Internal Report.* Fac. Civil Engineering, Delft University of Technology, Delft, The Netherlands, 1971.
8. De Hoop, A.T.: The Moving-Load Problem in Soil Dynamics - The Vertical Displacement Approximation. *Report Nr. ITS/EM 2001-04.* Department of Electrical Engineering, Delft University of Technology, Delft, The Netherlands, 2001.

Tortuosity Measurement and Its Spin-off to a New Electrical Conductivity Standard

JOS G.M. VAN DER GRINTEN
NMi Van Swinden Laboratory, Dordrecht, The Netherlands

The design of a new standard for the specific electrical conductivity of liquids is based on an instrument, called the Eindhoven conductivity cell, used to determine the tortuosity or mass-coupling factor of a porous structure. This paper describes the measurements that were performed during the design process and gives a schematic display of the designed standard.
From the measurements the following results were found. The specific electrical conductivity can be measured with an uncertainty of approximately 1%. The dominant uncertainty source are the distance between the electrodes, the temperature stability, the resistance measurement and the uncertainty contribution from the least squares fit. No significant frequency dependence of the specific conductivity was demonstrated. However, the uncertainty of the fit and the frequency of the bridge are positively correlated, which implies that DC operation results in the most accurate standard.

Introduction

Electrical conductivity measurements of liquids play an important role in analytical chemistry, in the process industry, in measurements of the environment and in other applications that require knowledge of the chemical concentration of solutions. The popularity of the conductivity measurements is the result of the simplicity and robustness of these measurement methods. The conductivity G is the reciprocal of the resistance R. The unit of the conductivity is the 'Siemens' (formerly called 'mho'). In a conductivity cell the conductivity G depends on the distance L between the electrodes, the area A of the electrodes and the specific conductivity σ of the liquid according to

$$G = \sigma A / L \qquad (1)$$

From this formula we find that the unit of the specific electrical conductivity σ is S/m (Siemens per meter).

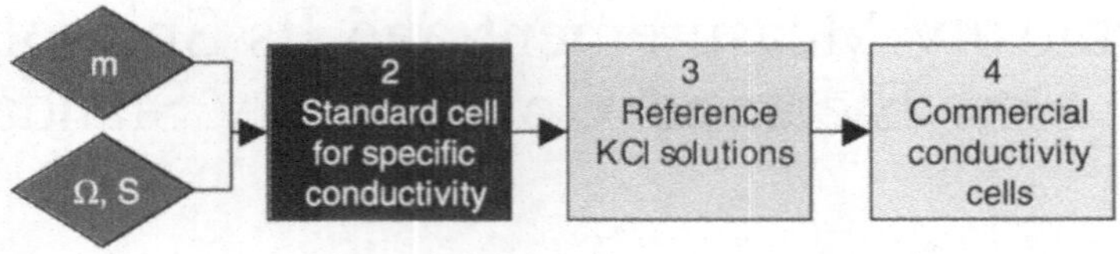

Figure 1. Traceability of conductivity cells.

The traceability of conductivity cells is shown in Figure 1. In order to calibrate commercially available conductivity cells (green box, 4), reference liquids (yellow box, 3) are used that are produced by creating a potassium chloride solution with a specific concentration [1]. Until the mid nineties these *KCl* solutions were the highest standards available to the industry.

In the mid nineties a number of laboratories like NIST and NPL started to develop absolute standards (blue box, 2) for the measurement of the specific conductivity of solutions. These standards are directly traceable to length and resistance units (red boxes on the left). In the same period started in NMi the development of a new conductivity cell based on a set-up used by Van der Grinten [2]. The objective of this development is to achieve formal traceability of these reference solutions and to increase the accuracy of conductivity measurements.

This paper describes the relevance of electrical conductivity for geodynamics, the experiments that were repeated at NMi in Delft and the subsequent design of a new standard. Tortuosity measurements and their geodynamic relevanceIn a fluid filled porous medium there are three interaction forces between the fluid and the porous material (Van der Grinten [2]). These are the fluid pressure on the porous structure, the drag forces described by the Darcy law or by the Forchheimer equation, and an unsteady contribution that can be expressed in terms in terms of the so-called added mass. The origin of this effect is that the pore fluid on pore scale is accelerated in a narrowing-widening structure and that the direction of the pore fluid acceleration on pore scale differs from the acceleration on continuum scale. The fluid density is apparently increased with

$$\rho_a = (\alpha\text{-}1)\ n\ \rho_f \tag{2}$$

in which α is the mass-coupling factor, also called tortuosity.

The mass-coupling factor is determined in an electrical conductivity experiment. This experiment is based on the analogy in the infinite frequency limit of the acceleration of an inviscid incompressible pore fluid in a rigid porous medium and the electrical conductivity of an electrolyte filled porous insulator. The mass-coupling factor is defined as the ratio of the apparent and the intrinsic fluid density. From the electrical analogy we find:

$$\alpha_\infty = n\ \sigma_f / \sigma_{eff} \tag{3}$$

where σ_f is the intrinsic fluid conductivity, σ_{eff} is the effective conductivity of the fluid in the porous medium, and n the porosity. The subscript 8 denotes the infinite frequency limit.

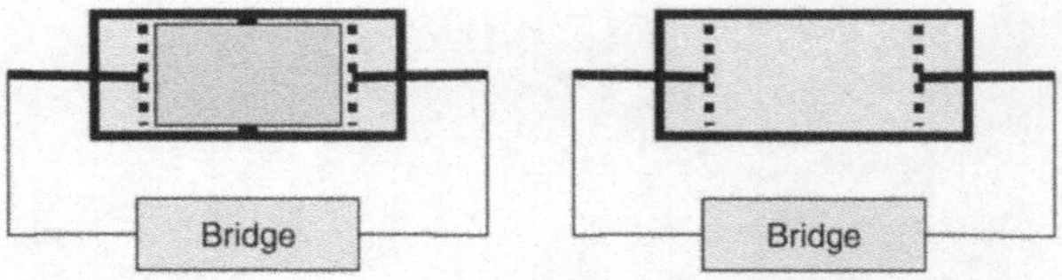

Figure 2. Set-up for measuring the specific electrical conductivity of a porous insulator saturated with an electrolyte (left) using a resistance bridge. On the right hand side the identical set-up without the porous medium.

The experimental set-up designed by Van der Grinten [2] is shown in Figure 2, left hand side. A coated porous cylinder is placed in a conductivity cell. The electrodes are placed against the porous solid. The porous medium is saturated with a de-aired 0,01 mol/l KCl solution. The conductivity is measured with a resistance bridge at 1000 Hz. The intrinsic fluid conductivity was measured in a cell of equal geometry without porous medium (Fig. 2). In order to eliminate all geometrical effects the conductivity versus the distance between the electrodes was measured.

Preliminary electrical conductivity measurements

When NMi started to develop a new electrical conductivity cell, the measurements carried out by Van der Grinten [2], were repeated in 1993 at NMi in Delft using identical equipment. The set-up was identical to figure 2, right hand side. In 1994 measurements were repeated once more using the Eindhoven cell, an *AC* impedance measurement bridge, Wayne Kerr model 6425, which splits the real and imaginary components of the impedance, a vernier caliper with an electronic display. Climate conditions were (23,5 ± 0,2) °C. The solution used was a KCl solution of 0,01 mol/l. The resistance measurements were carried out using a two-wire method and an excitation voltage of 1 V. The vernier caliper and the impedance bridge were both calibrated in advance, thus making the specific conductivity directly traceable to resistance and length standards.

The results of all measurements are graphically displayed in figure 3. The resistance is shown versus the distance of the electrodes. The graphs contain all data obtained in 1993 and 1994. The trend lines are obtained by a least squares approximation. The conductivity is calculated from the slope m of the trend line: $\sigma = 1/(A \cdot m)$. A review of the results is given in Table 1.

In the graph the trend lines of the 1994 measurements are parallel, showing a frequency dependent resistance shift. The trend line of the results of 1993 intersects the trend line of the results obtained at 1 kHz in 1994.

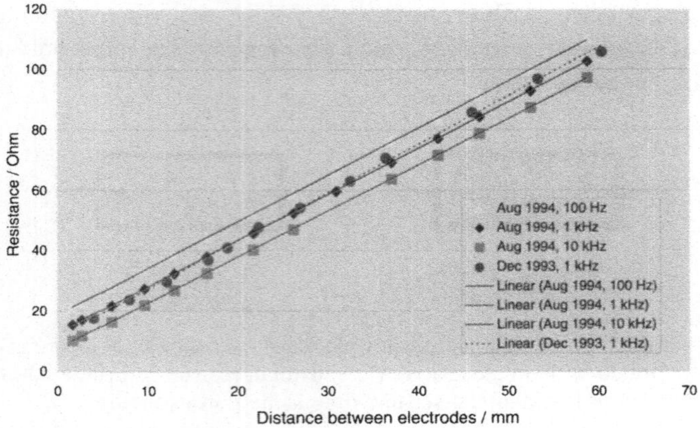

Figure 3. Results of the conductivity measurements. The resistance (Ω) is measured versus the distance between the electrodes (mm).

Table 1. Specific conductivity σ and corresponding uncertainty $u(\sigma)$ of a 0,01 mol/l *KCl* solution at (23,5 ± 0,2) °C. The literature value is interpolated from [1]. Its uncertainty is based on the uncertainty of the concentration of the KCl solution.

Date	f [Hz]	σ [mS/cm]	$u(\sigma)$ [mS/cm]	$u(\sigma) / \sigma$
Aug 1994	100	1,423	0,012	0,87%
Aug 1994	1000	1,442	0,014	0,94%
Aug 1994	10000	1,445	0,014	0,94%
Dec 1993	1000	1,375	0,030	2,2%
Literature [1]	-	1,366	0,021	1,5%

In conformity with the *GUM* [3] an uncertainty analysis was carried out. The results from this analysis for the measurements at 100 Hz are displayed in Table 2. The resulting uncertainties for the other measurements are summarized in Table 1. The uncertainties shown refer to the values for L and **R** that were measured in the center of the data points. All uncertainties are represented as the equivalent of double standard deviations. From Table 2 it appears that the most important uncertainty sources are the measurement of the distance between the electrodes, the temperature stability of the liquid, the resistance measurement and the fit of the data points.

Considering the measurement uncertainties the specific conductivities obtained at different frequencies, mutually agree. The result obtained in 1993 agrees well within uncertainty with the literature value. The difference between the results from 1993 and 1994 appears to be significant. A reason for this difference lies in the measurement method. The bridge used in 1993 measures the impedance in which both resistance and capacitance are included. On the other hand the bridge used in 1994 splits the resistance and capacitance, which leads to different results. The reference data in [1] assume that the full impedance is measured.

The regression analysis produced interesting results. The uncertainty of the fit of the specific conductivity in the 1994 measurements is one tenth of the 1993 results. When comparing the fits of the 1994 data it is observed that the uncertainty of the fit decreases with decreasing frequency. This implicates that *DC* conductivity measurements are the most accurate measurements. In addition the *DC* reference resistors have a lower uncertainty than the *AC* resistance references, which allows the realization of an even more accurate standard for the specific electrical conductivity.

Table 2. Overview of uncertainty analysis of the specific conductivity measurement of a 0,01 mol/l KCl solution at 100 Hz. The figures at bottom right hand side of the table are the root sum square summation of all the above uncertainty contributions.

Quantity x_i	Value	Sensitivity factor c_i		source	$u(x_i)$	$c_i \cdot u(x_i)$ [mS/cm]	$c_i \cdot u(x_i)/\sigma$
R (Ω)	45,809	σ/R	3,15E-05	1	0,023	7,2E-04	0,050%
			3,15E-05	2	0,005	1,7E-04	0,012%
L (cm)	2,511	σ/L	0,000576	1	0,017	1,0E-02	0,690%
			0,000576	3	0,010	5,8E-03	0,398%
d (cm)	7,600	$2\sigma/d$	0,00019	1	0,010	1,9E-03	0,132%
			0,00019	4	0,010	1,9E-03	0,132%
σ (mS/cm)	1,445	1	1	5	4,1E-03	4,1E-03	0,284%
			1	6	5,2E-03	5,2E-03	0,360%
σ (mS/cm)	1,445					1,4E-02	0,939%

Note. Specification of uncertainty sources: 1 calibration uncertainty, 2 repeatability of measurements, 3 electrodes not completely parallel, 4 un-roundness of the cell, 5 uncertainty of the slope of the least squares fit, 6 uncertainty due to a temperature dependency of the specific conductivity of 1,8%/°C and an uncertainty of 0,2 °C.

Towards a new standard

The objective of a new standard is to make the specific conductivity of *KCl* reference solutions directly traceable to resistance and length units. The insights obtained during the previously described measurements led to the design of new conductivity standard displayed in Figure 4. Two identical current sources provide a current through the resistors and the conductivity cell that is regarded as two variable resistors. By displacing the middle electrode the bridge is balanced. The specific electrical conductivity is derived from the position of the middle electrode. A more extensive description of the conductivity cell can be found in [4, 5].

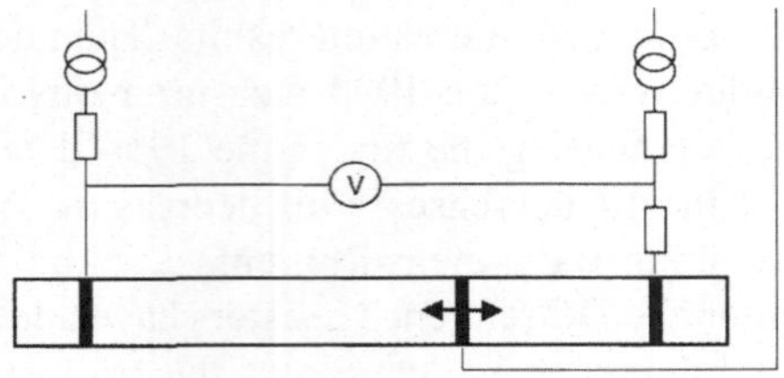

Figure 4. Schematic layout and electrical scheme of the NMi conductivity cell.

Conclusions

From the previous sections the following conclusions can be drawn.

The Eindhoven conductivity cell allows direct traceability of the specific electrical conductivity to resistance and length units. Experiments have led to the insights regarding the critical points in the design of a new standard.

For the measurement of the specific electrical conductivity the most important uncertainty sources are the distance between the electrodes, the temperature stability of the liquid and the fit of the data points.

From the uncertainty analysis of the linear regression it appears that frequency and uncertainty of the fit are positive correlated. This means that the most accurate results may be obtained in a *DC* operated cell. Another advantage is that the reference resistance at *DC* conditions has a lower uncertainty than the resistance at *AC* conditions.

A new cell design is based on the principle of the Wheatstone bridge. The objective of the new cell is to certify *KCl* reference solutions so that these solutions can be used for the calibration of commercially available conductivity cells.

References

1. OIML IR 56: *Standard solutions reproducing the conductivity of electrolytes*, 1981.
2. Grinten, J.G.M. van der: An experimental study of shock induced wave propagation in dry, water-saturated, and partially saturated porous media, PhD thesis Eindhoven University of Technology, 1987.
3. BIPM/IEC/IFCC/ISO/IUPAC/IUPAP/OIML: *Guide to the expression of uncertainty in measurement.* First edition, ISO 1993.
4. Patent NL C 1011540 (2000): *Meetcel voor het meten van de geleidbaarheid van vloeistoffen en werkwijze voor het meten van de geleidbaarheid van een vloeistof.*
5. Hafkenscheid, L.Th., Brinkmann, F,Trarbach A.A.M, de Leer E.W.B.: Development of a novel primary standard for measurement of conductivity, in: *Proceedings of the international Conference on Metrology.* Jerusalem, 2000, pp.86-91.

Simplified Models for Critical Speed of Moving Loads on Layered Halfspace

PAUL HÖLSCHER
GeoDelft, Delft, the Netherlands

A high speed train can go faster than the wavespeed in the soil under the track. In the Netherlands this problem is not imaginary. The wavespeed of the surface waves in the western part of the country is 40-80 m/s (= 150-300 km/h). This is surpassed by modern high speed trains. At this moment the trainspeed is up to 160 km/h, so the new trains are beyond practical experience. A trainspeed near the critical speed leads to a larger deflection and consequently to a loss of comfort for passengers and a strong need for suitable engineering works. Therefore, it is required to have some preliminary insight in the problem. The problem is studied theoretically by many researchers, amongst them at Delft University of Technology: Dieterman & Metrikine [3], Verruijt [8] and Verruijt & Cornejo Córdova [9].

In this article some simplified analytical solutions will be discussed. It turns out that a first global study of the phenomenon can be done without the usage of advanced numerical models. The primary idea is the following. If the critical speed of the soil with embankment is (much) higher than the trainspeed, no further research is needed. If the critical speed of the soil with embankment is not higher than the trainspeed, the dynamic deflection must be checked. If the design doesn't fulfill these conditions, it must be adjusted or checked with more advanced models.

In chapter 2 the first step, determination of the critical speed, is addressed. In chapter 3 the second step, determination of the dynamic multiplication factor is addressed.

General outline of analytical methods

In this article we are interested in the response of the surface of a horizontally layered halfspace which is loaded at its surface. For the problem with stationary load analytical methods are available. In general the following steps are required to generate an analytical solution.

- A Fourier transform is used to eliminate the dependency on time
- A Hankel transform is used to eliminate the dependency on horizontal distance.

The derivation used in this article for the axi-symmetric case is created by Van der Poel [6] and is based on Kennett [4]. This description uses generalized variables. It results in a description of the problem in the p-ω domain, where

p is the horizontal slowness, and ω the (angular) frequency. The boundary, interface and radiation conditions are transformed twice. The problem can be solved in this p-ω domain with the depth as the only independent variable.

The solution in time-space domain is calculated by applying inverse transformation. In the inverse Hankel transform one of the problems is related to the treatment of the poles. The location of the poles are defined by the slowness of the most important waves for the point considered. For a point at the surface it is the surface or Rayleighwave (see e.g. Achenbach [1], Van der Poel [6]). Since the inverse Hankel transformation is carried out in the frequency domain, the slowness of the Rayleighwave in the frequency domain is calculated. The (lowest) Rayleighwave speed can be used as an estimate for the critical speed in a layered halfspace.

Examples for critical speed

In this paragraph two examples are considered:
- a layered halfspace
- the Ledsgard case

Now a layered halfspace is considered. Table 1 presents the material properties. The properties of the top-layer are based on sand, the intermediate layer on clay/peat and the bottom halfspace on sand. The choices are based on correlation with results of cone penetration test and properties are representative for a location along the railway track Amsterdam-Utrecht.

Table 1. Material properties for example Amsterdam-Utrecht

Property	Embankment	Holocene	Pleistocene
volumetric mass [kg/m^3]	1800	1400	2000
bulkmodulus [MPa]	871	50.4	1150
shearmodulus [MPa]	87.1	5	115
damping [%]	0.68	2	1.44
thickness [m]	1.8	5	

Figure 1 shows the calculated Rayleighwave speed. For the lowest frequencies, the Rayleighwave speed is determined by the properties of the Pleistocene sand. For the highest frequencies the Rayleighwave speed is determined by the properties of the embankment. In the intermediate frequencies, the properties of the soft Holocene layer is important. The thickness of the layers determines the "depth" of the curve in Figure 1. As can be seen from Figure 1 the chosen depth of 2 m for the embankment leads to an estimate for the critical speed of 88 m/s. Taking into account the uncertainty in the parameter estimation and a possible thicker Holocene layer, the 2 m embankment might be too low for a trainspeed of 200 km/h (= 56 m/s), but is far too low for a trainspeed of 300 km/h (= 83 m/s).

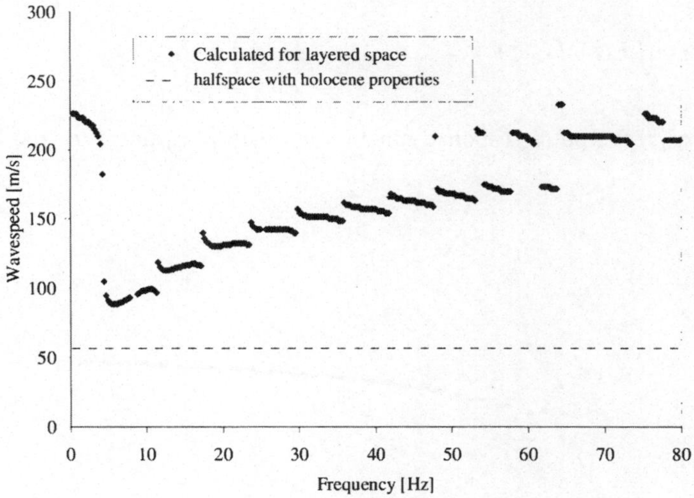

Figure 1. Rayleighwave speed on frequency for Amsterdam-Utrecht case.

As a second example the Ledsgard case (described by Andreasson, Bengtson & Zackrisson [2]) is calculated. The soil parameters are chosen from Madshus & Kaynia [5]. Table 2 shows the soil parameters used. Figure 2 shows the dispersion curve. The lowest value of the Rayleighwave speed is 51 m/s which is quite close to the value estimated from measurements and which is indeed a lower limit.

Table 2. Material properties for Ledsgard case (from Madshus & Kaynia [5])

Property	Embankment	clay-1	clay-2a	clay-2b
volumetric mass [kg/m3]	1835	1325	1427	1529
bulkmodulus [MPa]	87.8	25.4	33.8	117.5
shearmodulus [MPa]	40.5	2.6	3.5	12.2
damping [%]	2.0	2.0	2.0	2.0
thickness [m]	1.4	3.0	8.0	

The deflection under the load

Once the estimated critical speed of a location is known and close to the design speed for the trains, it is important to calculate the deflection under the train speed. A simplified method might lead to a first estimate.

An approximate solution can be derived by the summation of fundamental solutions. Start with the deformation u of the surface by a time dependent load $F(t)$. From convolution it follows

$$u(y_0,t) = \int_{-\infty}^{t} F(\tau) h(y_0,\tau) d\tau$$

with $h(y_0,t)$ the pulse response function at position y_0 and $F(t)$ the load.

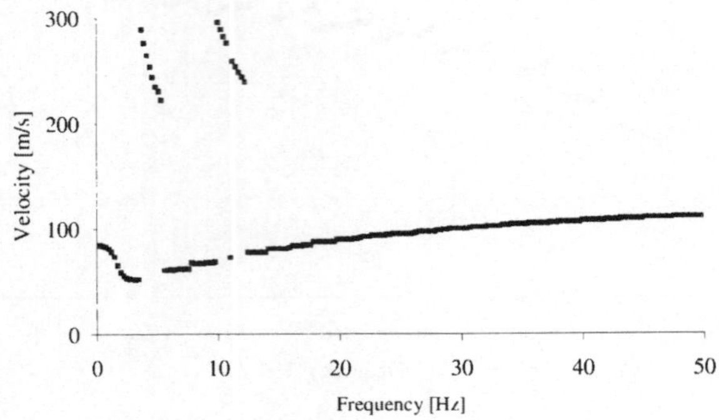

Figure 2. Rayleighwave speed on frequency for Ledsgard case.

For a line load on the surface the equation is extended

$$u(y_0,t) = \int_{-\infty}^{\infty} \left\{ \int_{-\infty}^{t} \tilde{F}(y,\tau) h(y_0,y,\tau) d\tau \right\} dy$$

with $\tilde{F}$ the load per unit length which equals the moving load.

$$\tilde{F}(y,\tau) = F_0 \, \delta(y - v\tau)$$

Now, the integral over y is immediate discretised by a summation. The load is distributed over the two nearest points by a distribution function

$$F_r = \frac{1}{2} F_0 \left(1 - \cos 2\pi \frac{y}{2L} \right)$$

$$F_l = F_0 - F_r$$

With L the distance between the nodes and y the position of the load, subscripts l and r refers to the left and right node of the interval. Thus the solution can be rewritten by

$$u(y_0,t) = \sum_{j=-N}^{N} \int_{-\infty}^{t} F_{nod}(\tau) h(y_0 - y_j, \tau) d\tau$$

$$\text{with } F_{nod}(\tau) = \frac{1}{2} F_0 \left(1 - \cos 2\pi \frac{v\tau}{L} \right) \text{ if } \tau < \frac{2L}{v}, = 0 \text{ elsewhere}$$

Since the integral with convolution can be calculated by an analytical solution, the final solution is just a summation of a number of such solutions.

Example for the deflection under a load

The simplified integral method for the dynamic deflection is tested for the case described above. The deflection is calculated for the Amsterdam-Utrecht case. A track-length of 32 m is taken into account, with the point of interest in the middle, discretised by points at 80 cm distance.

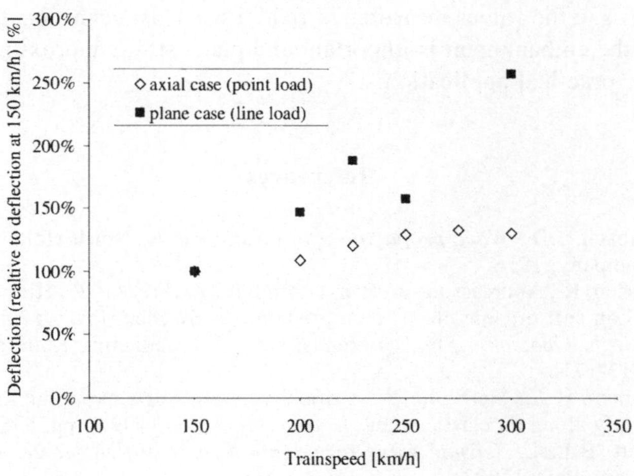

Figure 3. Amplification of deflection with speed Amsterdam-Utrecht case.

Two cases are considered: the case of a circular loaded area moving over the surface and the case of a lineload (perpendicular to the direction of motion) moving at the surface. For the axial case the vertical force is distributed over a circular area with radius of 40 cm. Since the absolute values of the deflection is very sensitive for the area of distribution, this value should be constant. Therefore, the results can only be used for relative calculations, thus comparing the deflections for several trainspeeds. The deflection under a standing load can be calculated using advanced static methods, so this isn't a limitation. For the lineload similar solutions are used, where the assumed $1/\sqrt{r}$ is removed artificially. Of course better results will be obtained by a correct plane strain approximation.

Figure 3 shows the maximum deflection on trainspeed. For the axial case the influence of train speed turns out to be limited, up to 33%. The plane strain case leads to a trainspeed dependent increase of the deflection of 150%, so at 300 km/h the deflection is 2.5 times the deflection at 150 km/h. The results are not consistent with regard to velocity, presumable due to the correction from axial case to plane case. The calculated increase of deflection in the plane strain case is well comparable with the measured deflection described by Woldringh & New [7].

Conclusion

This article shows that it is possible to assess the critical trainspeed problem with relatively simple models, which are directly based on analytical solutions for time-dependent stationary loads. The dispersion curve in the frequency domain gives a good estimate of the critical speed in layered media. The integration of the surface deflection due to time dependent stationary loads shows the effect of critical speed. For plane strain the speed dependent increase fits to the values measured in field tests. This means that the limited width of the embankment is important and plane strain approximations are useful for practical application.

References

1 Achenbach, J.D.: *Wave propagation in elastic solids*. North Holland Publishing Company, 1976.

2 Adolfsson, K., Andréasson, B., Bengtsson, P.E., Zackrisson, P.: High speed train X2000 on soft organic clay – measurements in Sweden, *Proc. 12th Eur. conf. Soil Mech. Geotechn. Eng.*, (Barends, e.a., ed.), Balkema, Rotterdam, 1999, pp. 1713-1718.

3 Dieterman, H.A., Metrikine, A.: Critical velocities of a harmonic load moving uniformly along an elastic layer, *J. Appl. Mech.*, 64 (1997), pp. 596-600.

4 Kennett, B.L.N., *Seismic wave propagation in stratified media*, Cambridge University Press, 1983.

5 Madshus, C., Kaynia, A.M.: Dynamic ground interaction, a critical issue for high speed train lines on soft soil, *Proc. 12th Eur. conf. Soil Mech. Geotechn. Eng.*, (Barends, e.a., ed.), Balkema, Rotterdam, 1999, pp.1829-1836.

6 Poel, T.R. van der: *The elastodynamic wavefield in horizontally layered media*. Msc. thesis, TUDelft, Faculty of Technical Mathematics, 1995.

7 Woldringh, R.F., New, B.M.: Embankment design for high speed trains on soft soils, *Proc. 12th Eur. conf. Soil Mech. Geotechn. Eng.*, (Barends, e.a., ed.), Balkema, Rotterdam, 1999, pp. 1703-1712.

8 Verruijt, A.: Dynamics of soils with hysteretic damping, *Proc. 12th Eur. conf. Soil Mech. Geotechn. Eng.*, (Barends, e.a., ed.), Balkema, Rotterdam, 1999, pp. 3-14.

9 Verruijt, A. & Cornejo Cordova, C.: Moving Loads on an Elastic Half-Plane With Hysteretic Damping, *ASME J. Appl. Mech.*, Vol. 68 (2001), pp. 915-925.

Poroelasticity and Biot Coefficients

DAVID M.J. SMEULDERS
TUDelft, Delft, the Netherlands

For poroelastic media, we are facing the problem of identifying the relevant parameters. Whereas the porosity appears quite naturally as the void fraction of the medium, for the stresses and pressures within a porous medium many definitions and approaches are used. This already goes back to the tragic controversy between the Viennese professors Paul Fillunger and Karl von Terzaghi in 1936/1937 as described by de Boer [1]. For wave propagation in porous media, the first comprehensive theory was established by Biot [2] and independently by de Josselin de Jong [3], although Frenkel [4] already in 1944 discussed rotational and dilatational waves in soil. So-called "gedanken" experiments on porous samples were used by Biot & Willis [5] to relate the parameters of the Biot theory to the compressibilities of the solid and fluid constituents. In this article it is shown that these relations can elegantly be obtained by straightforward application of continuity and constitutive laws following Verruijt [6].

Pressure and stresses

Considering a fluid-filled elastic skeleton with a statistical distribution of interconnected pores, the porosity is usually denoted by:

$$\phi = \frac{V_\phi}{V_b}, \tag{1}$$

where V_ϕ is the volume of the pores contained in a sample of bulk volume V_b. It is understood that the term "porosity" refers to the effective porosity, thus including only the interconnected void spaces as opposed to those pores which are sealed off. In the following, the word "pore" will refer to the effective pores, while the sealed pores will be considered as part of the solid volume V_s. The total stress tensor in the bulk material is:

$$\begin{pmatrix} \tau_{11}+\tau & \tau_{12} & \tau_{13} \\ \tau_{21} & \tau_{22}+\tau & \tau_{23} \\ \tau_{31} & \tau_{32} & \tau_{33}+\tau \end{pmatrix}, \tag{2}$$

the fluid in the pores we may write:

$$\tau = -\phi p. \tag{3}$$

The remaining components τ_{ij} of the total stress tensor are the *tension* forces per unit *bulk* area applied to that portion of the cube faces occupied by the solid. They are a result both of the fluid pressure p and the additional intergranular stresses σ_{ij}:

$$\tau_{ij} = -\sigma_{ij} - (1-\phi)p\delta_{ij}, \tag{4}$$

where the Kronecker symbol δ_{ij} is obviously introduced because the pore fluid cannot exert nor sustain any shear forces. We notice that the additional intergranular stresses σ_{ij} are defined negative in tension. They are called "additional" because they add up to the fluid pressure induced stresses in the solid (see (4)). We may also define the tension forces per unit *solid* area A_s applied to that portion of the cube faces occupied by the solid:

$$p_{s_{ij}} = \tau_{ij}A_b/A_s = -\sigma_{ij}/(1-\phi) - p\delta_{ij} \tag{5}$$

Obviously, the total normal tension force per unit *fluid* area A_f applied to the fluid part of the faces of the cube can be written as $\tau A_b/A_f = -p$. From (3) and (4) the total stress tensor in the bulk material may also be written:

$$\begin{pmatrix} -\sigma_{11} - p & -\sigma_{12} & -\sigma_{13} \\ -\sigma_{21} & -\sigma_{22} - p & -\sigma_{23} \\ -\sigma_{31} & -\sigma_{32} & -\sigma_{33} - p \end{pmatrix} \tag{6}$$

This expression for the total stress tensor is also given by Verruijt ([6], equation 1), where it must be noticed that he has denoted the total stress tensor σ_{ij} and the intergranular stress $\bar{\sigma}_{ij}$. We will now consider two commonly used tests in which the volume effects caused by pore pressure and intergranular stresses can be studied separately.

Unjacketed test

When a porous sample is fully submerged in a water tank (pressure change dp_e) and the sample is assumed to be fully water-saturated, it is immediately clear that the fluid pressure must be continuous over the interface:

$$dp_e = dp. \tag{7}$$

For the intergranular stresses at the interface we may write:

$$d\sigma_{11} = d\sigma_{22} = d\sigma_{33} = 0 \tag{8}$$

As there are no changes in the intergranular stresses, the unjacketed test is used to study the volume effects caused by the pore pressure changes. Defining the matrix bulk modulus K_a, the bulk volume change $dV_b = -K_a^{-1}V_b dp$ is measured in this test. In case of homogeneous bodies, whether or not isotropic, the application of an incremental pressure dp_e means applying this increment both to the outer and inner pore surface, which leads to a linear mapping and does not change the porosity ϕ ($d\phi = 0$). Therefore we may write for the volume change of the matrix grains:

$$dV_s = (1-\phi)dV_b = -\frac{1}{K_a}V_s dp. \tag{9}$$

This means that for homogeneous media K_a can also be interpreted as the bulk modulus of the single grains.

In case of non-homogeneous bodies, it can no longer be argued that the porosity is not influenced by a pressure change. Instead, it seems appropriate to introduce a bulk modulus of the pore volume $K_\phi^{-1} = -(\phi V_b)^{-1}\partial(\phi V_b)/\partial p$, which leads to

$$d\phi = -\left(\frac{\phi}{K_\phi} - \frac{\phi}{K_a}\right)dp. \tag{10}$$

From the relation $d\phi = d[1 - V_s/V_b]$ it can be found that

$$dV_s = -\frac{1}{(1-\phi)}\left(\frac{1}{K_a} - \frac{\phi}{K_\phi}\right)V_s dp. \tag{11}$$

Though it is straightforward to determine the coefficient K_a by measuring the displacements of the boundary surfaces of the sample, it is not immediately clear how to determine K_ϕ. This problem was discussed by Biot & Willis [5] who proposed measuring the increment of pore fluid content.

Jacketed tests

In this case, a porous sample is jacketed and fully submerged in a water tank (pressure change dp_e) and the inside of the jacket is made to communicate with the atmosphere through a tube to ensure constant internal fluid pressure. We may write (see (6) and $dp = 0$):

$$dp_e = d\sigma_{11} = d\sigma_{22} = d\sigma_{33} \tag{12}$$

As there are no pore pressure changes, the jacketed test is used to study the volume effects caused by intergranular stresses. Defining the matrix bulk modulus K_b, the bulk volume change $dV_b = -K_b^{-1}V_b d\sigma$ is measured in this test, where σ is the isotropic component of the intergranular stress: $\sigma = \sigma_{kk}/3$. In the literature, it is often assumed that a dry specimen exhibits the same properties as a fully saturated one and therefore the conventional jacketed test is usually performed on a dry specimen. Assuming that the response of the solid particles to a unit increase of the average stress induced by the intergranular forces equals the response to a unit increase of the uniform stress induced in these particles by the fluid pressure, we may write for the volume change of the particles (see (5) and (9)):

$$dV_s = -\frac{1}{(1-\phi)}\frac{1}{K_a}V_s d\sigma \tag{13}$$

This relation between the particle volume and the intergranular stress is also described by Verruijt ([6], equation 3), where he initially assumes that the volume change is not described by $[K_a(1-\phi)]^{-1}$, but by a separate parameter β_p. Again using the relation $d\phi = d[1 - V_s/V_b]$, we can find for the porosity change:

$$d\phi = -\left(\frac{1-\phi}{K_b} - \frac{1}{K_a}\right) d\sigma \tag{14}$$

Apparently, there is always a porosity change in these type of compression tests, whether or not we are discussing homogeneous samples. It can also be argued that a small increase of the intergranular stress must result in a decrease of the porosity, so $\partial\phi/\partial\sigma < 0$. From (14), we then find that $(1-\phi)K_a > K_b$, which was also previously stated by Verruijt ([6], equation 7).

Stress-strain relations

In the entire fluid-solid system chemical reactions are assumed not to occur. The solid skeleton is considered to have compressibility and shearing rigidity, and the fluid may be compressible. The deformation of a unit cube is assumed to be completely reversible and elastically linear. Each volume element is described by the averaged displacement of its fluid $\mathbf{U}(\mathbf{r}, t)$ and of its solid parts $\mathbf{u}(\mathbf{r}, t)$. The strain components for the solid and the fluid, respectively, are:

$$e_{ij} = \frac{1}{2}\left(\frac{\partial u_j}{\partial x_i} + \frac{\partial u_i}{\partial x_j}\right) \tag{15}$$

$$\varepsilon_{ij} = \frac{1}{2}\left(\frac{\partial U_j}{\partial x_i} + \frac{\partial U_i}{\partial x_j}\right). \tag{16}$$

The bulk volume change can be described as a function of both the pore pressure change and the change of the intergranular stress and thus as a summation of the effects in jacketed and unjacketed tests:

$$dV_b = -\frac{1}{K_b} V_b d\sigma - \frac{1}{K_a} V_b dp \tag{17}$$

This relation may be rewritten (see [6], equation 33):

$$-\frac{\partial\sigma}{\partial t} = K_b \frac{\partial e}{\partial t} + \frac{K_b}{K_a}\frac{\partial p}{\partial t}, \tag{18}$$

where $e = e_{kk} = dV_b/V_b$. A generalization now seems appropriate. From (18) it can easily be seen that the stress-strain relation for the porous system may be written according to Verruijt ([6], equations 1 and 32):

$$-\sigma_{ij} = (K_b - \frac{2}{3}G)e\delta_{ij} + 2Ge_{ij} + \frac{K_b}{K_a} p\delta_{ij}. \tag{19}$$

In the literature, the effective stress σ'_{ij} is often introduced in such a way that the deformation of the matrix is fully determined by that stress (see [6], equations 1 and 14):

$$-\sigma'_{ij} = -\sigma_{ij} - \frac{K_b}{K_a} p\delta_{ij} = (K_b - \frac{2}{3}G)e\delta_{ij} + 2Ge_{ij}. \tag{20}$$

Continuity and constitutive equations

The change in the particle volume can be described as a function of both the pore pressure change and the change of the intergranular stress and thus as a summation of the effects in jacketed and unjacketed tests. Considering a homogeneous body, we refer to (9) and (13). For the fluid, the bulk modulus K_f is introduced. The constitutive relations then become:

$$\frac{1}{\rho_s}\frac{\partial \rho_s}{\partial t} = \frac{1}{K_a}\frac{\partial p}{\partial t} + \frac{1}{(1-\phi)}\frac{1}{K_a}\frac{\partial \sigma}{\partial t} \tag{21}$$

$$\frac{1}{\rho_f}\frac{\partial \rho_f}{\partial t} = \frac{1}{K_f}\frac{\partial p}{\partial t} \tag{22}$$

The linearized continuity equations are:

$$\frac{\partial}{\partial t}[(1-\phi)\rho_s] + (1-\phi)\rho_s \nabla \cdot \mathbf{v} = 0, \tag{23}$$

$$\frac{\partial}{\partial t}(\phi\rho_f) + \phi\rho_f \nabla \cdot \mathbf{w} = 0 \tag{24}$$

where we have defined the averaged velocities $\mathbf{w} = \frac{\partial \mathbf{U}}{\partial t}$ and $\mathbf{v} = \frac{\partial \mathbf{u}}{\partial t}$ of the fluid and solid parts of an arbitrary volume element. It is noticed that we have eleven unknown variables: ρ_s, ρ_f, p, σ, ϕ, $\mathbf{v}$ and $\mathbf{w}$. Because they are vectors, the velocities $\mathbf{v}$ and $\mathbf{w}$ count as three unknowns each. As we have only formulated five equations (continuity and constitutive equations and the stress-strain relation (18)), we will not be able, so far, to solve the system completely. The system is closed by the momentum equations for the solid and the fluid introduced by Biot [2]. These momentum equations will not be discussed here. From the combination of the solid relations (21) and (23) and the fluid equations (22) and (24) respectively, we may write:

$$\frac{1-\phi}{K_a}\frac{\partial p}{\partial t} + \frac{1}{K_a}\frac{\partial \sigma}{\partial t} - \frac{\partial \phi}{\partial t} + (1-\phi)\nabla \cdot \mathbf{v} = 0 \tag{25}$$

$$\frac{\phi}{K_f}\frac{\partial p}{\partial t} + \frac{\partial \phi}{\partial t} + \phi \nabla \cdot \mathbf{w} = 0. \tag{26}$$

Elimination of the porosity term by adding the equations yields:

$$\left(\frac{1-\phi}{K_a} + \frac{\phi}{K_f}\right)\frac{\partial p}{\partial t} + \frac{1}{K_a}\frac{\partial \sigma}{\partial t} + (1-\phi)\nabla \cdot \mathbf{v} + \phi \nabla \cdot \mathbf{w} = 0. \tag{27}$$

This relation is usually called the storage equation, which forms a basic relationship in consolidation problems (see [6], equation 24).

We may now eliminate either σ or p from the storage equation (27) with (18) and through use of the identity $\partial e/\partial t = \nabla \cdot \mathbf{v}$. When σ is eliminated this yields:

$$\beta \frac{\partial p}{\partial t} + K_f \left(1 - \phi - \frac{K_b}{K_a}\right) \nabla \cdot \mathbf{v} + \phi K_f \nabla \cdot \mathbf{w} = 0, \tag{28}$$

where we have introduced $\beta = \phi + K_f/K_a(1 - \phi - K_b/K_a)$. Rewriting (28) as

$$-\phi \frac{\partial p}{\partial t} = \phi \frac{K_f}{\beta} \left(1 - \phi - \frac{K_b}{K_a}\right) \nabla \cdot \mathbf{v} + \phi^2 \frac{K_f}{\beta} \nabla \cdot \mathbf{w}, \tag{29}$$

it is now immediately possible to recognize the Biot coefficients Q and R, which are defined according to

$$-\phi \frac{\partial p}{\partial t} = Q \nabla \cdot \mathbf{v} + R \nabla \cdot \mathbf{w} \tag{30}$$

In a similar way, it is possible to recombine (19) and (28) to obtain the other Biot equation

$$-\frac{\partial \sigma_{ij}}{\partial t} - (1 - \phi) \frac{\partial p}{\partial t} \delta_{ij} = A \nabla \cdot \mathbf{v} \delta_{ij} + 2G e_{ij} + Q \nabla \cdot \mathbf{w} \delta_{ij}, \tag{31}$$

where $A = \phi K_b/\beta + (1 - \phi) K_f (1 - \phi - K_b/K_a)/\beta - 2/3G$. The Biot parameters A, Q, and R are now uniquely related to measurable properties of the poroelastic material, which allows for a comprehensive study of wave phenomena in porous media by investigating the dependence of the Biot parameters as a function of the poroelastic properties.

References

1. de Boer, R.: *Theory of Porous Media.* Springer, 2000.
2. Biot, M.A.: Theory of propagation of elastic waves in a fluid-saturated solid. *J. Acous. Soc. Am.* 28 (1956), pp.168-191.
3. de Josselin de Jong, C.: What happens in soil during pile driving? *De ingenieur* 68 (1956), pp.B77-B88.
4. Frenkel, I.: On the theory of seismic and seismoelectric phenomena in a moist soil. *J. Phys. (USSR)* 8 (1944), pp.230-241.
5. Biot, M.A. & Willis, D.G.: The elastic coefficients of the theory of consolodation. *J. Appl. Mech.* 24 (1957), pp.594-601.
6. Verruijt, A.: The Theory of Consolidation. In: *Proc. NATO Advanced Study Institute on Mechanics of Fluids in Porous Media.* Nijhoff, 1982, pp.349-368.

Experiences in Wave Propagation and Vibration by Moving Loads

HERKE G. STUIT
Holland Railconsult, Utrecht, the Netherlands

The extension of high-speed train connections through densely populated urban areas and between city centers together with the increase of train speed require not only control of the rail-track dynamics but also the propagation of vibrations in the subsoil and into building. Wave propagation and vibration reduction are recently described and treated by theoreticians and are measured at several locations in Europe. This paper will give an overview of several locations where track displacements are a risk. Based upon the measurements finite element models have been used to compute track displacements of high-speed trains. The computations are used to design track sections with vibration reduction measures to diminish dynamic amplification of a track.

The critical train speed

The phenomenon of a train travelling at the critical train speed is comparable to a plane the flying through the sound barrier. In both cases the vehicle continuously generates waves that are travelling away at a certain velocity. If the vehicle velocity is lower than the wave propagation, the emitted waves are simply travelling away from the source. When the vehicle velocity is equal to wave propagation the wave fronts emitted in the direction of the vehicle movement show a concentration of the arrival times of successive waves, causing large displacements of the track. This velocity is the so-called critical velocity. Various authors have studied this phenomenon (Dieterman [2], Heelis [4], Kenney [5], Verruijt [8]).

When a solid medium is dynamically loaded, several types of waves will travel from the source point through the medium at different velocities. The three important ones are: the pressure wave, the shear wave and at the free surface the surface (Rayleigh) wave. Their velocities are defined as

$$c_p = \sqrt{\frac{G}{\rho}\frac{2(1-\nu)}{(1-2\nu)}} \; ; \qquad c_s = \sqrt{\frac{G}{\rho}} \quad ; \qquad c_R = 0.9 * c_s \tag{1}$$

Where G is the shear modulus, ν the Poisson's ratio and ρ the density. The Rayleigh wave approximately contains about 2/3 of the total seismic energy and has a relative geometric damping. Therefore, the Rayleigh wave is the most important seismic wave for a critical train speed.

In reality, there are different soil layers and various track embankments, generating different waves. Beforehand, it is not obvious which type of wave will be responsible for the critical train speed. When minimum wave propagation is much higher than the train speed, one does not have to worry about large dynamic amplifications. On the other hand, the opposite does not hold. It is possible that no large dynamic amplifications occur, even if one of the velocities described above would be lower than the train speed.

Experiences

Measurements of track displacements relative to the train which indicate a dynamic amplification of the track, have recently been reported by several authors (Eijgenraam [3], Madshus [6], Woldringh [9]). Figure 1 shows the measurements at different locations. All measurements show a significant increase of the displacements with an increasing speed.

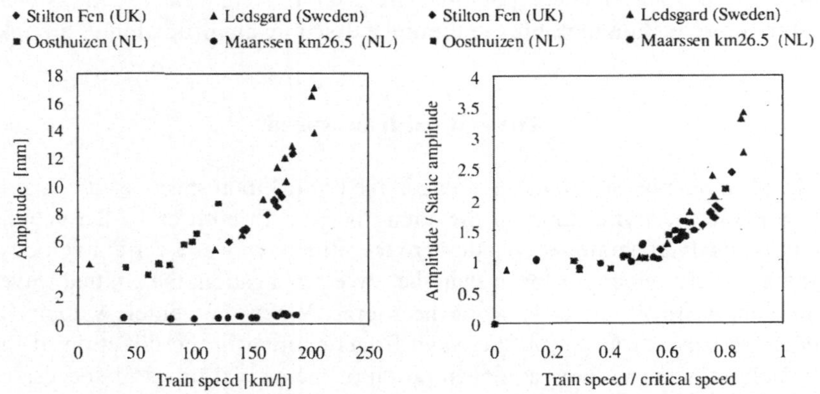

Figure 1. Measurements Stilton Fen (UK), Ledsgard (Sweden), Oosthuizen (NL) and Maarssen (NL). Peak to peak displacement versus train speed (a); relative displacement versus train speed relative to the critical speed (b).

In Figure 1b the displacement is displayed as a function of the velocity of the moving load relative to the estimated overall critical velocity. The displacements are divided by the static displacement, so dynamic amplification factors (ratio of maximum displacement and static displacement) can be compared. Figure 1b shows the similarity between various test sites in the UK, Sweden and the Netherlands. Dynamic amplification factor vary from 2 to 4.

Modeling

To model the critical train speed linearly elastic soil behaviour is assumed. Furthermore, it is assumed that the cross-area of the track bed and the surrounding soil remains the same along the railway line. This means that the response of a complete travelling train can be modeled by summing the response of a series of single pulses. The response of a linear system of differential equations due to a pulse is called the Green's function. If we look at the displacements along the track and assume that the track is parallel to the x-axis the Green's function can be expressed as

$$G(x,t;x_s,\tau) \tag{2}$$

By convoluting the Green's function with a source time function $g(x_s,\tau)$ with a certain moving velocity, the response of an arbitrary source time function can be modeled. The following equation holds for the displacement at location x and time t

$$u(x,t) = \int_{-\infty}^{\infty} G(x - v\tau, t - \tau; 0,0)\, d\tau \tag{3}$$

This equation represents the vertical displacement along the track due to a moving load. The Green's function is the vertical response along the track due to a point source excitation that happens at time t=0 and at location x=0.

Green's functions exist in analytical form for a layered and homogeneous half space. For a more complex geometries, for instance, with a track bed structure, the analytical form of the Green's function becomes complicated. In this case the embankment and the soil layers are modeled with a 3D finite element model. It has the advantage that the critical train speed of any track bed geometry can be modeled. The finite element model should automatically contain all types of waves, so one does not have to worry about which type of wave causes the critical train speed.

Figure 2. Finite element model to compute the Green's function.

Figure 2 shows an example of the Holland Railconsult finite element model which resembles a double track on an infinite layered half space. The model is loaded with a short pulse distributed over the width of the track at the position where symmetry boundary conditions apply. At the bottom and the right hand side, non-reflecting boundary conditions are applied, since reflections at the boundaries distorts the results.

Computations Amsterdam – Utrecht railway

Dutch Rail intends to upgrade the travelling speed from 140 km/h to 200 km/h at the railway Amsterdam-Utrecht. Since a preliminary geotechnical survey indicated weak soil layers. A full-scale test campaign was set up focussing at 4 locations with vertical displacement transducers at the top of the embankment. The measurements were carried out with a Thalys train travelling up to 200 km/h. In order to determine which sections of the railway would have to be improved, finite element computations have been made for the locations km 16.9 and km 26.6. The parameters used in the finite element analyses are given in Branderhorst [1] and Eijgenraam [3].

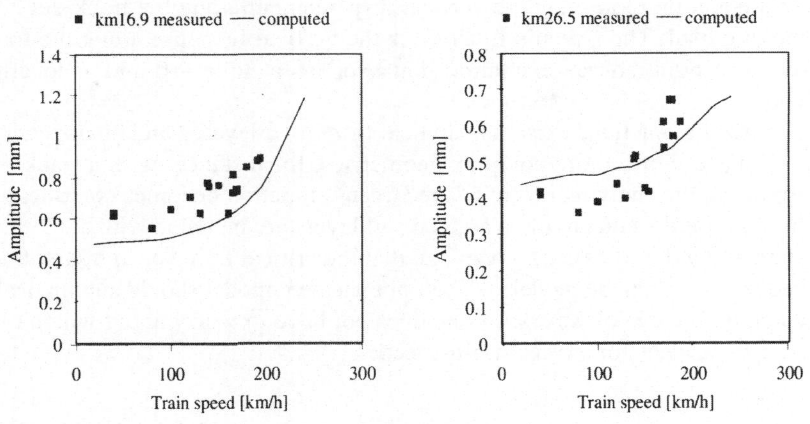

Figure 3. Computed results for 2 locations at the railway Amsterdam-Utrecht (NL). Peak to peak measured displacement versus train speed compared with measurements. Location km 16.6 (a); location km 26.5 (b).

In Figure 3 measured displacements are compared with computed displacements at different velocities. The variation of the measurements is quite large. Therefore, the displacements of individual measurements for one velocity can not be compared to one individual computation. However, at km 26.5 and km 16.9 measured displacements are comparable to computed values.

The determination of material parameters using shear wave velocity measurements is very sensitive. Displacements are approximately linearly dependent on the inverse of the stiffness of the soil layers, and therefore, proportional to the square of the inverse of the shear wave velocity c_s. The shear wave velocity c_s and Rayleigh wave velocity c_R will hardly be influenced by slightly different material parameters. Therefore, the error in the prediction of the critical train speed will be relative less than the error of the predicted displacements with slightly different material parameters.

Both predictions and calculations show that the critical train speed will not be reached at the maximum train speed on the track Amsterdam–Utrecht and the maximum displacements will stay below the limit of 3 mm.

Reduction measures

In Table 1 several vibration reduction measures for railways are listed. Depending on the situation one of the measures is usually best applicable and most effective. The high-speed line Amsterdam-Paris is designed as a continuous piled viaduct. In the test site of No-Recess (Woldringh [9]) several methods of ground improvement have been examined and at Oosthuizen (Stuit [7]) the effect of a concrete slab foundation has been examined.

Table 1. Vibration reduction measures

Train	Reduction train weight; axle arrangement; wheel-flat control.
Track	Smoothing track surface (irregular control); softening track support; stiffening track bending rigidity.
Embankment	More massive and rigid structure; wave impeding barrier; floating track; piled viaduct.
Subsoil	Ground improvement.

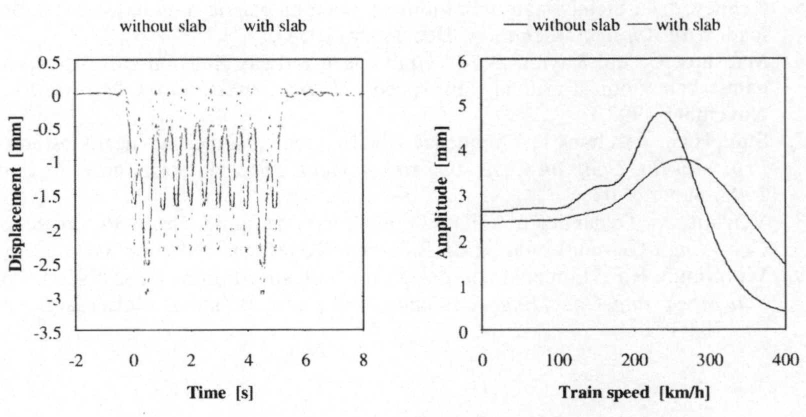

Figure 4. Computed results of the effect of a concrete slab. Displacement of the track due to a Thalys train at 140 km/h (a); displacement versus train speed (b).

For location km 16.9 at the line Amsterdam-Utrecht the effect of a concrete slab in the track bed has been examined. Figure 4 shows it for a 0.45 m thick concrete slab at this location. The amplitude of the displacement in Figure 3 is smaller than the amplitude in Figure 4. The reason is that the displacement transducers were not placed directly under the track, but in the maintenance path approximately 2 meters away from the track. The critical velocity increases from 244 to 263 m/s. The most important effect of the slab is the reduction of the dynamic amplification and the absolute values of the displacements. A slab reduces the dynamic amplification factor from 1.82 to 1.57. Absolute values of the displacements due to the passage of a Thalys high-speed train are displayed in Figure 4a. The Thalys is a high-speed train comparable to the French TGV. At the front and the back of the train a motor car is connected to a passenger car, which are the only positions where two bogies are located close to each other. The maximum displacement occurs at the locations where these two bogies pass.

References

1. Branderhorst, J.F.J. *Modelleren voor het boeggolfprobleem bij hogesnelheids-treinen*. M-Sc Thesis, Holland Railconsult, Utrecht, 1997.
2. Dieterman, H.A. and Metrikine, A.V.: Steady-state displacements of a beam on an elastic half-space due to a uniformly moving constant load. European. *J. of Mechanics of Solids*, Vol 16(2) (1997), pp.295-306.
3. Eijgenraam, A.A., Gardien, W. and Stuit, H.G.: Measured vertical displacements of the railway Amsterdam-Utrecht (The Netherlands) versus different model computations. *Proc. Int Conf Soil Mech and Geot. Eng.* (Togrol et al., ed.), Instanbul, 2001, (to be published).
4. Heelis M.E., Collop A.C., Dawson A.R., Chapman D.N., Krylov V.V.: Transport effects of highs speed trains crossing soft soil. In Barends et. al. (eds): *Proc. Eur. Conf. Soil Mech. and Geot. Eng.*, (Barends et at., ed.), Balkema, Rotterdam, 1999, pp.1809-1814.
5. Kenney, J.T.: Steady-state vibrations of beam on elastic foundation for moving load. *J. of Applied Mechanics,* December (1954).
6. Madshus, C. and Kaynia, A.M.: High speed railway lines on soft ground: dynamic behaviour at critical train speed.. *6th Int. Workshop on Railway Noise,* November 1998.
7. Stuit, H.G.: Test track low frequency vibration reducing measure near Oosthuizen. *Proc. 4th Int. Conf. on Case Histories in Geot. Eng.*, (Prakash, ed.), St. Louis, 1998, paper: 4.03.
8. Verruijt, A.: Dynamics of soils with hysteretic damping. *Proc. Soil Mech. and Geot. Eng.*, (Barends et al., ed.), Balkema, Rotterdam, 1999, pp.3-14.
9. Woldringh, R.F.: Embankment design for high speed trains on soft soils. *Proc. Soil Mech. and Geot. Eng.,* (Barends et al., ed.), Balkema, Rotterdam, 1999, pp.1703-1712.

GEOSTATICS

Unexpected Phenomena in Geotechnical Centrifuge Tests

HENDERIKUS G.B. ALLERSMA
University of Delft, Delft, the Netherlands

In the course of the past decade, a large number of small-scale tests have been carried out in the small geotechnical centrifuge of the University of Delft, on a variety of geotechnical problems. In the centrifuge testing technique, the soil samples are subjected to stress conditions similar to those experienced at full scale. This guarantees a more realistic behavior of the small models. In model tests complicated problems can be analyzed and new ideas can be tested easily. An advantage of model tests is that the behavior of a structure can be monitored much more accurately than in a full-scale test. A small size centrifuge allows several tests to be performed in quick succession. This is particularly effective for the study of failure mechanisms and the influence of changes in design. Furthermore, it is possible to prepare small soil samples with precision, so that the effect of slight differences in design can be made visible. This has facilitated the observation of several unexpected phenomena, which are not yet predicted by mathematical models. Aspects of the following subjects will be discussed in this paper: behavior of conical footings; bearing capacity of suction piles; widening of road embankments; cratering during gas blowout; slope stability and failure of sand embankments during water infiltration.

Test facilities

The centrifuge of the University of Delft

The geotechnical centrifuge at the University of Delft [1] was designed at the Geotechnical Laboratory of the Department of Civil Engineering and Geosciences. It is a beam type centrifuge, with two swinging carriers (Fig.1a). A 1500 mm long beam is connected to the shaft, allowing it to be rotated in the horizontal plane. The two swinging carriers are connected to the beam by means of brackets. Each carrier consists of two plates held apart at a distance of 410 mm by four connecting beams. The surface of a plate is 400 x 300 mm. Samples with a weight of 400N can be accelerated up to 300 g (12g-tons).

The system electronics enable the performance of computer-controlled tests in flight (Fig.1b). To minimize electrical disturbances, the control unit is placed in the spinning part of the centrifuge. The unit contains a small single board PC compatible computer, a 12-bit analog to digital converter

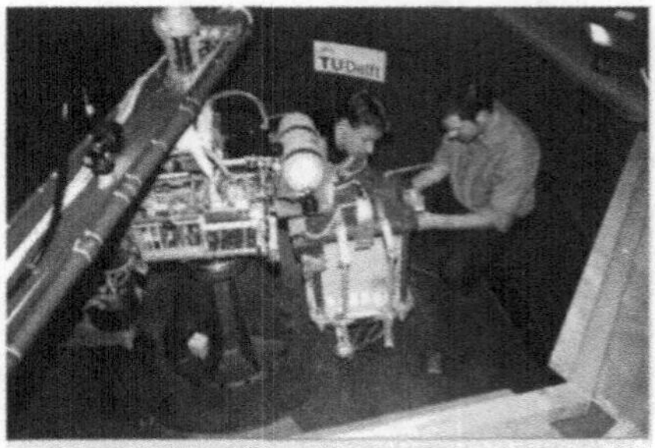

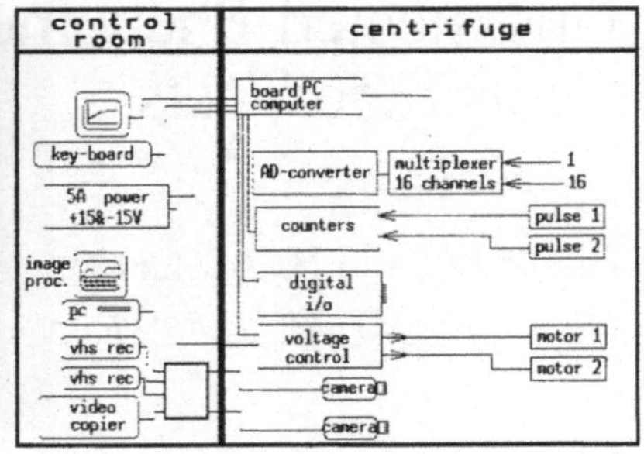

a b

Figure 1. The small centrifuge of the University of Delft (a). Electronic system (b).

with a 16-channel multiplexer, two voltage controlled outputs of 8 amps each; two 16-bit counters and several digital inputs and outputs. A 1Gbyte hard disk is located at the precise center of the centrifuge, where it operates correctly up to at least 160g at 1 meter. During a test, the relevant parameters are displayed in graphical form and stored on the hard disk. A special feature is that several phenomena can be measured using the video images. In this technique the video images of the in-flight test are captured by the frame grabber in the PC and processed until the relevant parameters are isolated and digitized. Image processing can be used to visualize and digitize the surface deformation of clay and sand samples, to digitize the consolidation of a clay layer or to digitize the displacement of objects [2,3]. The use of light ensures that the measurement does not influence the test in any way.

Devices for in flight tests

Several devices have been developed for the purpose of performing advanced tests in flight [4]. The two-dimensional loading system (Fig. 2a) can be considered to be a universal tool, with application in several tests. Adjustable vertical and horizontal displacements of 100mm and 200mm, respectively, can be achieved with an accuracy of better than 0.1mm. The loads are measured in the two perpendicular directions and used in a computer program to control the device. The system is able to apply loads in excess of 5 kN. In addition to loading tests, the device can be used as a robot to manipulate tests in flight.

An in-flight sand pouring machine (Fig.2b) with a width of 150mm is available to construct embankments in flight. The sand can be poured at a particular location or layer by layer over 150mm. The strategy can be defined in a computer program. The device can be used to simulate the widening of embankments.

Some tests require gas to be supplied to a soil sample. The gas has to be stored in the spinning section of the centrifuge. To make the storage as compact as possible, two high pressure (200 bar) 5 liter cylinders are mounted on the centrifuge beam. A computer-controlled system adjusts the pressure and the gas flow. Flow rates of 10 l/s can be achieved. The gas in the high pressure

a b

Figure 2. two dimensional loading system (a), sand raining machine (b).

cylinders represents a considerable amount of power, which can be used, in principle, for tests calling for large loads or energy. The gas supply system is used for several tests, such as pile driving, water circulation, control trapdoor test, simulating gas blowouts in soil layers and simulation of a suction pile installation. Other tools available to perform a large variety of in-flight tests include a pile driving hammer, in-flight excavation, wave simulator and pollution infiltration.

Sample preparation

An important aspect of centrifuge research is sample preparation. Comparison of the results of different tests is possible only if good reproducibility of the samples is achieved. An automated computer-controlled curtain rainer has been developed to prepare well defined sand layers in the test containers [3] The falling height of the sand can be adjusted in order to control the density. The height is kept constant during raining by means of an optical sensor and an actuator. Sand samples with a surface area of 300x300 mm and a maximum thickness of 150 mm can be prepared. The porosity can be varied between 35% and 39%. The sand samples can be reproduced with a standard deviation of less than 0.2 percent. The preparation of a sample takes about 20 minutes.

The preparation of clay samples is dependent on the type of test that has to be performed. In many cases pottery clay can be used. The small size of the models means that commercially available quantities are sufficient. In tests requiring normally consolidated clay, a slurry with a water content of more than 100% is consolidated at the desired g-level. This assures the most realistic water content and strength over the height. A special centrifuge is available for this task.

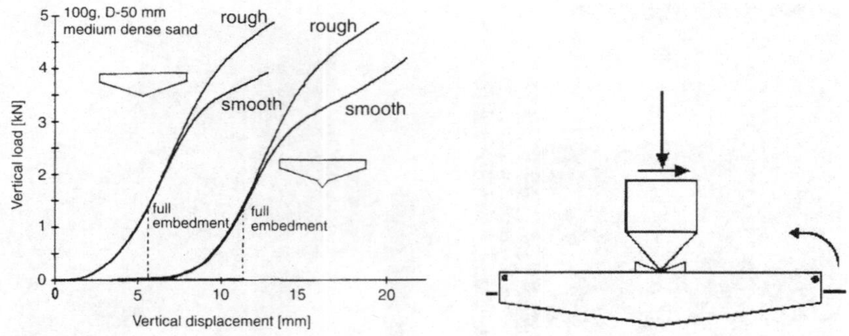

a b

Figure 3. Behavior of spudcans during vertical (a) and horizontal (b) loading.

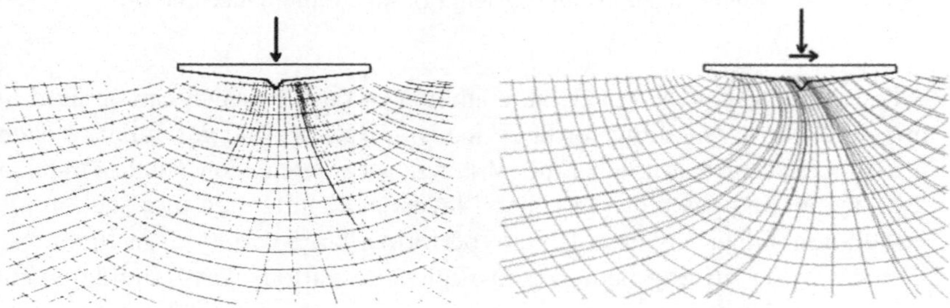

a b

Figure 4. Optically measured principal stress trajectories. Vertical loading (a), Combined horizontal and vertical loading (b). (5 meter prototype at 100g).

Test results

Behavior of conical footings

The research on the behavior of conical footings is related to the foundation of mobile offshore structures. The main goal was to examine the sliding behavior of shallow footings on sand, by simulating the loading path which is assumed to occur under storm conditions [5]. The test results have been used to examine the reliability of the design rules. A number of interesting phenomena could be observed in the test program on medium dense sand. Figure 3a shows a plot of the response of vertically loaded spudcans at 100g, where footings are used with and without tip and with different surface roughness. The non-linear course of the first part of the curve clearly shows the influence of the conical shape of the footing. It appeared that roughness had no significant influence on the course in the first part. The final bearing capacity was not reached at full embedment. An extra displacement of about 0.05 times the diameter was required. The roughness of the footing was varied by using a smooth metal surface and a surface of sandpaper. The graphs show that the roughness significantly affects the vertical bearing capacity (more than 20%). It was found that the horizontal sliding resistance of a rough cone

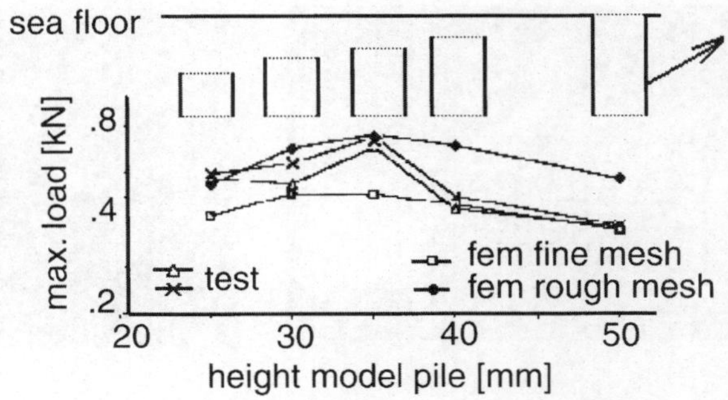

Figure 5. Horizontal bearing load of suction piles, installed below sea floor.

was 28% more than a smooth spudcan. Increasing roughness seems to be a cheap way to improve the bearing capacity of spudcans.

The influence of the tip is also visible. In the case of smooth spudcans, the bearing capacity of footing with tip is 14% lower than the footing without tip. In the tests with the rough surface the difference is 7%. Apparently the tip disturbs the soil somewhat, resulting in a lower vertical bearing capacity. On the other hand, the tip increases the horizontal bearing capacity. In the case of a smooth or rough cone an increase of approximately 25% could be observed [6].

In Figure 3b, a horizontal loading test was performed on a vertically loaded spudcan, where a free rotation of the footing was possible. A surprising phenomenon was that the footing rotates counter clockwise. In Figure 4, the principal stress trajectories are made visible in a centrifuge test by using optically sensitive granular material [7,8]. It can be seen that the stress rotates as a result of the horizontal components (Fig.4b). The stress rotation indicates that a counter clockwise rotation of the footing is to be expected.

Bearing capacity of suction piles

Suction pile foundations are recognized as an effective method for anchoring floating devices and to fix offshore structures to the sea floor. The installation of suction piles is attractive, because only a pump is required to penetrate a pile with a height of 10 m and a diameter of 9 m in a few hours time into the sea floor.

The anchor is often loaded in the horizontal direction. A test program has been carried out [9] to examine the horizontal bearing load in relation to several conditions, such as the loading angle, pile height and diameter.

The need to save material, led to the invention of a pile whose top lies below the sea floor (Fig. 5). In soft clay in particular, the upper part of the pile does not contribute much to the bearing capacity. The installation should occur by a pile segment with a dome, which can be recovered after the installation process. In order to examine the effect of this installation technique, a series of centrifuge tests were carried out on piles installed at the same depth,

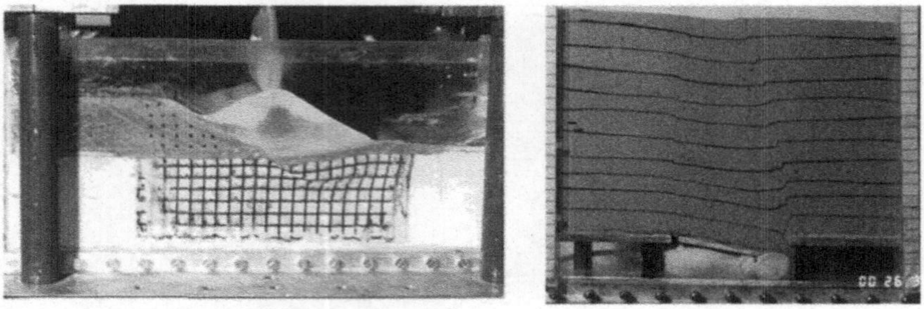

a b

Figure 6. Widening of embankment on clay (a). Simulation of a displacement gradient in sand (b).

but with different height. An unexpected result was that the horizontal bearing capacity increases with decreasing length of the piles. If the height of the pile was reduced by 30%, the bearing capacity increases by approximately 25%. The effect was less strong in clay, but increased bearing capacity could be observed nonetheless. It was found that the bearing capacity was also larger in the case of pure vertical pullout load. The reason for the increasing bearing load is probably that, in the case of a full size pile, friction occurs along the pile wall. If the pile is below sea floor, friction is generated in the soil, apparently resulting in larger pullout loads.

Widening of a road embankment

Several tests have been carried out to examine the behavior of existing sand embankments on soft soil during its widening [10]. The aim of the research program was to examine the mechanism that causes cracks in the asphalt pavement. The extent to which the sand supply scheme influences the cracking behavior of the asphalt layer was also investigated. A typical set of test results is shown in Figure 6a. The deformation of the clay after widening is made visible by grid lines, where markers are used to make the behavior of the existing embankment visible. It was clearly visible that there were no horizontal displacements. This means that this cannot be the reason for cracks in the asphalt.

However, a gradient in the vertical displacement could be observed underneath the embankment. Thanks to this gradient, the embankment is subjected to bending, which causes tensile stresses in the asphalt cover. It can be deduced that gradient of only a few degrees can cause cracks of 100 to 200mm. Vertical displacement of the asphalt cover were also observed in some cases in the field. This observation could give the impression that a more complicated deformation mechanism is active in the clay layer. In Figure 6b, however, it is shown that abrupt vertical displacement in a sand layer is possible at some distance from the source, even when the source causes only a gradient in vertical displacement.

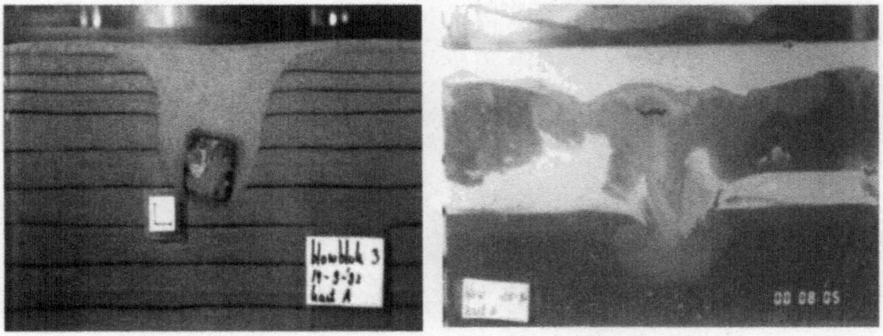

a b

Figure 7. Examples of gas blowouts sinking material (a); clay layer in sand (b)

Cratering during gas blowout

A gas blowout can occur during the drilling process in the oil industry. Fracturing can allow a loss of the mud used to keep the pressure of the reservoir under control. A mixture of gas and oil can reach the surface, causing a crater of fluidized soil (Fig.7). This may result in loss of equipment (Fig.7a), or in triggering a chain reaction where there are other wells in the vicinity. In order to form a view on the depth of the crater, gas blowouts were simulated in sand layers at 150g in the centrifuge [11]. The gas was introduced at the bottom of the sample. It appeared that a rather sudden start was required to create a crater at all. However, if an element was placed at the surface, a crater was formed under conditions of a gradual increase in gas flow. The tendency was observed in the field that the crater bottom did not pass clay layers. However, in centrifuge tests it was observed that a clay layer causes a crater. The gas accumulates in the first instance under the clay layer, causing a sudden supply when it breaks through.

Failure of sand embankments during water infiltration

In order to gain a better insight into the failure behavior of embankments during wave overtopping, several test programs have been carried out in the centrifuge [12]. A model of a rather common construction of an embankment used for water protection is shown in Figure 8a. The embankment consists of a sand body which is covered with a clay layer. Water infiltration occurs through cracks in the clay layer during wave overtopping. In the centrifuge test at 80g, water is infiltrated at the crest of the dike. The two-dimensional model in combination with transparent boundaries allowed the rise of the groundwater table to be observed. The stream lines have been made visible by means of a tracer.

A surprising observation was that no shear band failure mechanism could be observed. Should the embankment be constructed of pure sand, failure starts at the toe, which is visible as gradual erosion as a result of seepage. If the sand embankment is covered with a clay layer, the increasing pressure during water table rising tilts up the clay cover (see stream lines in Figure 8a). This

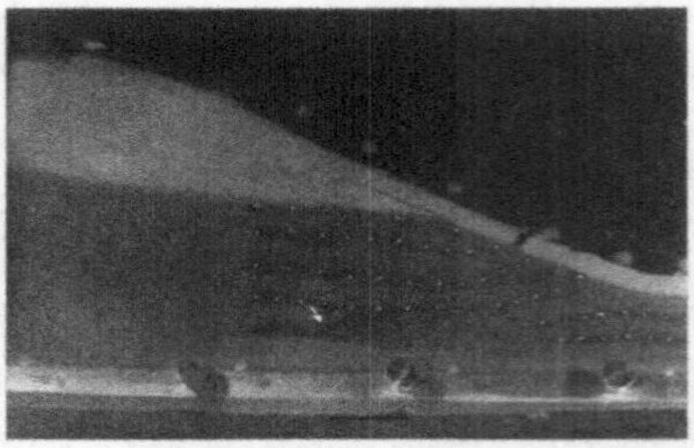

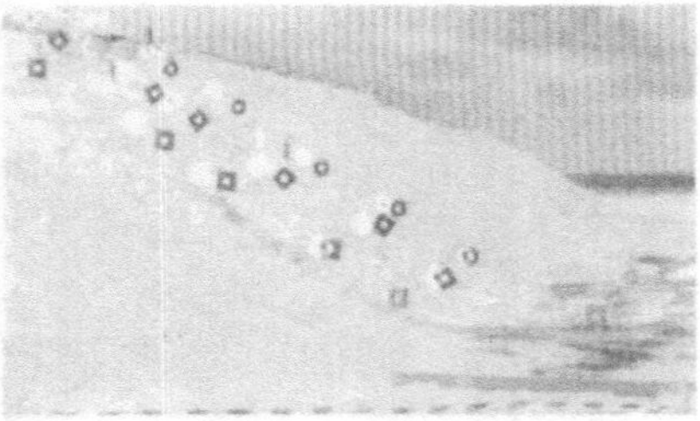

a b

Figure 8. Simulation of embankment failure centrifuge test (a); field test (b).

causes friction reduction resulting in cracking of the clay layer under its own weight. The location of the crack appeared to be dependent on the thickness of the clay layer. The larger the thickness, the higher the crack. Once the crack is there, water will erode the embankment more easily.

Figure 8b shows the result of a field test on an embankment construction similar to that used in the centrifuge. In this test wave overtopping was simulated by running water over the slope surface. The surface deformation was visualized by labels at the slope surface and by subtracting exposures of different stages.

It was also visible in this test that the motion of the clay layer is mainly parallel to the slope surface. In this case, the crack was close to the crest, where some buckling can be observed at the toe of the embankment (caused by water pressure). Afterwards, a cut was made in the embankment in order to examine whether shear band failure could be visualized. However, even with the help of clay columns, no shear bands could be identified.

Conclusions

A number of tests have been carried out in the small centrifuge in order to examine a variety of problems. The small centrifuge has been found to be very convenient for trial and error tests. The small models could be modified easily, and several tests could be performed in a short period. Thanks to this, several different conditions could be examined in a modest amount of time, and at a very reasonable cost. In several cases, surprising phenomena have been observed, which could not easily be deduced from exercises with mathematical models. On the other hand the centrifuge tests were helpful in validating analytical and numerical calculations.

The centrifuge tests were helpful in understanding the soil and soil-structure behavior. In several cases only centrifuge tests were able to get a better insight into the problem. Some test results are used directly in engineering practice and for improving the monitoring strategy in field tests.

Acknowledgement

Many thanks are devoted to the technicians of the laboratory, Mr. A. Mensinga, Mr. J. van Leeuwen and Mr. J.J. de Visser.

References

1. Allersma, H.G.B.: The University of Delft geotechnical centrifuge. Proc. Int. Conf. Centrifuge94, (1994), pp. 47-52.
2. Allersma, H.G.B.: On line measurement of soil deformation in centrifuge tests by image processing. Proc. Int. Conf. on Experimental Mech., (1990), pp. 1739-1748.
3. Allersma, H.G.B., H.G. Stuit, P. Hölscher: Using image processing in soil mechanics. Proc. XIII Int. Conf. on Soil Mech. and Found. Eng., (1994), pp. 1341-1344.
4. Allersma, H.G.B.: Development of miniature equipment for a small geotechnical centrifuge. Transportation Research Record No. 1432, (1994), pp. 99-105.
5. Allersma, H.G.B., B. Hospers, J.G. den Braber: Centrifuge tests on the sliding behaviour of spudcans. Canadian Geotechnical J., 34(5) (1997), pp. 658-663.
6. Allersma, H.G.B.: Centrifuge tests on the influence of shape on the sliding behaviour of spudcans. Proc. Int. Conf. BOSS, (1997), pp.225-233.
7. Allersma, H.G.B.: Optical analysis of stress and strain in photoelastic particle assemblies, Thesis, Delft University of Technology, Delft, the Netherlands (1987).
8. Allersma, H.G.B.: Stress analysis on photoelastic particle assemblies in centrifuge tests. Int. conference Centrifuge98, (1998), pp.61-67.
9. Allersma, H.G.B., A.A. Kirstein, R.B.J. Brinkgreve, T.Simon: Centrifuge and numerical modelling of horizontally loaded suction piles. 9th Offshore and Polar Eng. Conference, (1999), pp.711-717.
10. Allersma, H.G.B., L. Ravenswaay and E. Vos. Investigation of road widening on soft soils using a small geotechnical., Transp. Research Record No.1462, (1994), pp. 47-53.
11. Allersma, H.G.B., W.J. van Niekerk, A.P. Kooijman: Simulation of cratering in a small geotechnical centrifuge. Int. Conf. Centrifuge94, (1994), pp. 325-330.
12. Allersma, H.G.B.: Centrifuge tests on slope failure during water infiltration. Int. Symp on Slope Stability Engineering (IS-SHIKOKU'99), (1999), pp..447-452.

Discrete Element Analysis of Tunnel Stability in Blocky Rock Masses

STEFAN VAN BAARS
TUDelft, Delft, the Netherlands & University of Sydney, Sydney, Australia

Most complex problems in soil mechanics are solved nowadays with finite element modelling (FEM). For some particular problems FEM is not the best solution for obtaining insight. For example, if we want to know more about a granular material and its relation between micro and macro behaviour, micro modelling with elements which can create and break its contacts, is required. This type of modelling is called discrete element modelling (DEM). Inspired by the work on DEM by Cundall and Strack (1979), the author developed in 1993 a computer model to describe the behaviour of granular materials. This model calculates the displacements of each individual grain and all forces on this grain, by using the equations of equilibrium. The correspondence between the numerical results on one hand and reality and analytical results on the other hand, was sufficiently good to conclude that the DEM is a valid tool for fundamental research.

At the University of Sydney the stability of underground excavations is subject of research. Model tests were carried out to get insight in the failure mechanism of the roof of underground openings in jointed rock. Unfortunately FEM and DEM could not predict the right failure load for these tests. Since DEM has helped solving previous discrete problems, the idea was raised to solve this problem with an improved discrete element model.

Discrete element modelling of blocky materials

In discrete element modelling, which is also called distinct element modelling, the displacement and deformation of an element depends only on the contact forces between this element and its neighbours. All elements will be displaced in such a way that a force equilibrium is found. The relative displacement between two elements can be described by three parameters: the displacement in normal direction n, the displacement in shear (tangential) direction s and the rotation p (figure 1). These displacements cause contact stresses, which can be resolved into a normal force F_n, a shear (tangential) force F_s and a moment M_p (figure 2). The normal force and the shear force are oblique forces because there is an angle β between the shear force and the contact surface at the interface of the two elements.

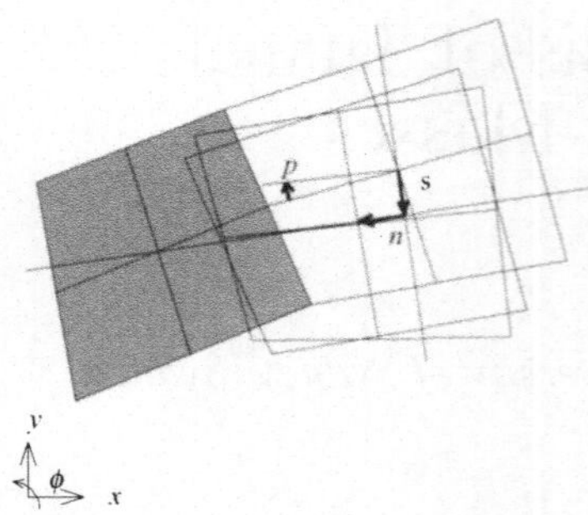

Figure 1. Contact overlap

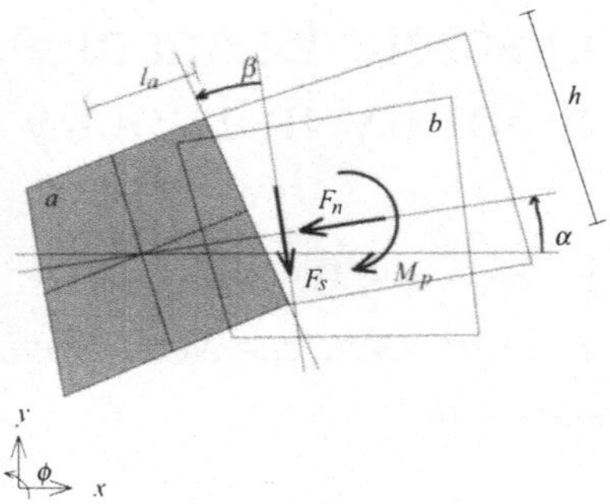

Figure 2. Contact forces

If the lateral contraction can be neglected and the elements are rather homogeneous and isotropic, we may assume a linear relation between the forces and the displacements

$$F_n = k_n \cdot n \quad , \quad F_s = k_s \cdot s \quad \text{and} \quad M_p = k_p \cdot p \tag{1}$$

The stiffnesses k_n, k_s and k_p depend on the Young's modulus E , the Poisson's ratio ν, and the shape of the element. In most cases the stiffnesses hardly influence the failure behaviour, so an approximation is sufficient.

The displacement of an element will be such that the variation of the forces defines an equilibrium state of this element

$$\sum \vec{F}_{old} + \sum \Delta \vec{F} = \sum \vec{F}_{new} = \vec{0} \tag{2}$$

For a 2-D solution, there are three equilibrium equations

$$\begin{bmatrix} \Delta f_{x,x} & \Delta f_{x,y} & \Delta f_{x,\omega} \\ \Delta f_{y,x} & \Delta f_{y,y} & \Delta f_{y,\omega} \\ \Delta f_{\omega,x} & \Delta f_{\omega,y} & \Delta f_{\omega,\omega} \end{bmatrix} \cdot \begin{bmatrix} \Delta x \\ \Delta y \\ \Delta \phi \end{bmatrix} = \begin{bmatrix} f_x \\ f_y \\ f_\omega \end{bmatrix} \tag{3}$$

With a simple summation over all the contacts of the element, all pivots in the previous matrix can be calculated. The new equilibrium position of the element can be directly calculated with Gauss elimination. Although this element is in equilibrium now, the displaced element will effect its neighbour elements. Therefore several iterations through the whole structure are necessary to find the total equilibrium state of all the elements.

Due to these displacements some contacts between the elements can break and therefore the contact forces are limited. Especially when there is no cohesion between the elements or the cohesion is less than the required stress. In both cases, the forces and moments are limited. Three situations are distinguished. A) The normal force must be positive because tension is not possible. B) The shear force is limited by a combination of the angle of internal friction ϕ and the angle of the contact surface β because of slip. C)

The ratio between the moment and the normal force on an element reaches its maximum when the stress becomes zero on one side of the element. If the elements are homogeneous and isotropic (linear stress distribution), the ratio will be 1/6 of the height of the elements (figure 3). The corresponding conditions are expressed as follows.

A) $0 \leq F_n$ (4a)

B) $\tan(\beta - \phi) F_n \leq F_s \leq \tan(\beta + \phi) F_n$ (4b)

C) $-\frac{1}{6} h F_n \leq M_p \leq \frac{1}{6} h F_n$ (4c)

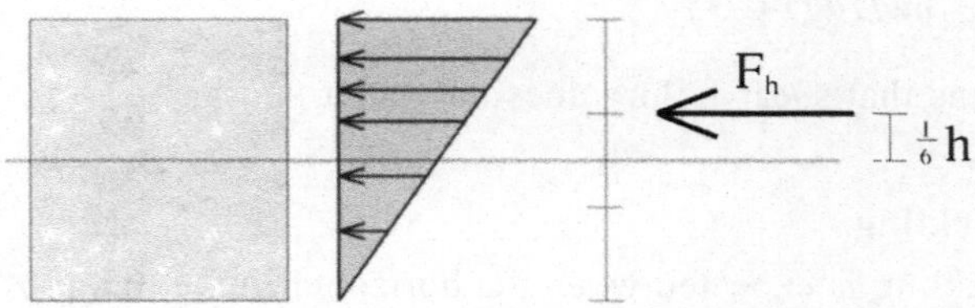

Figure 3. Maximum ratio between moment and normal force

Openings in rock

Model tests

In order to gain some insight in the failure mechanism and the moment of failure of underground openings in jointed rock, Mackenzie (1996) and Shepherd (1996) have used a scale model with 20 × 30 square polyacetal blocks representing rocks (γ = 13.6 kN/m^3, E = 4400 MPa and tan(ϕ) = 0.21). One of their excavations is schematised, see figure 4. The roof of the opening started to buckle at a (flexible) vertical load of $\sigma_v = \sigma_1$ = 100 kPa when the (rigid) horizontal load was held constant at $\sigma_h = \sigma_3$ = 100 kPa.

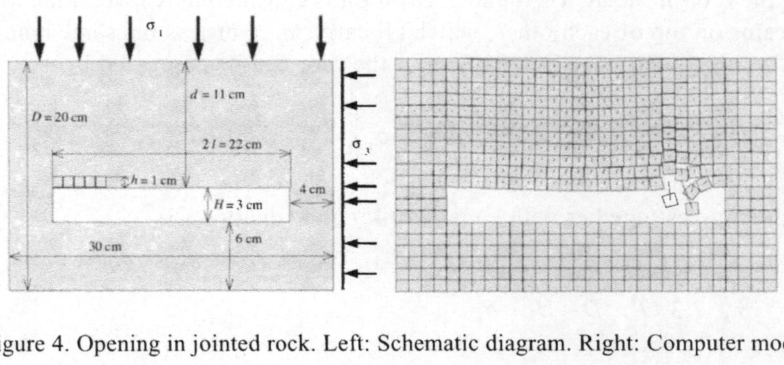

Figure 4. Opening in jointed rock. Left: Schematic diagram. Right: Computer model

Simulations with a DEM computer model show failure at the same loading pressure as the scale model. Since there is no crushing of the blocks, the question now arises, if failure of the excavation in the scale model was caused by shearing, elastic buckling, tensile buckling or something else.

Shear failure

Shear failure is expected when friction between elements is too low to carry the vertical load, so $F_v/F_h > \tan(\phi) = 0.21$. If the distribution of the horizontal stresses is more or less linear, then the estimated average horizontal and vertical forces on the elements above the corners of the roof are

$$\frac{F_v}{F_h} = \frac{\sigma_v hwl/d}{\sigma_v hwD/(D-H)} = 0.11 < 0.21 \tag{5}$$

which means that shear failure does not occur yet.

Elastic buckling

Elastic buckling is expected when the horizontal force in a roof element is

$$F_h = EI\frac{\pi^2}{l^2} \quad \text{with:} \quad I = \frac{1}{12}wh^3 \tag{6}$$

Combined with the horizontal force of equation (5), this becomes

$$E = \sigma_h \frac{12}{\pi^2}\left(\frac{l}{h}\right)^2 \frac{D}{D-H} = 132 \text{ MPa} \tag{7}$$

In our case, the normalised Young's modulus should be at most 132 MPa, but because the stiffness of the blocks is more than 30 times higher, elastic buckling will not take place.

Tensile buckling

If the roof of the underground excavation is considered as several identical beams on top of each other, which all carry more or less the same amount of vertical load, then the moment in the roof can be expressed by

$$M = \frac{1}{8}ql^2 \quad \text{with:} \quad q = w\left(\gamma h + \frac{h}{d}\sigma_v\right) \tag{8}$$

which gives together with equation (4c) and equation (5)

$$\frac{\sigma_v}{\sigma_h} = \frac{4}{3}\cdot\frac{dh}{l^2}\cdot\frac{D}{D-H} - \frac{d\gamma}{\sigma_h} \tag{9}$$

With this solution, the vertical load for failure should be σ_v = 108 kPa in the present case, which is close to both the measured and numerical calculated σ_v of 100 kPa. This analytical solution is based on several assumptions of the blocks above the excavation, like:

1. The blocks are homogeneous and isotropic.
2. The stress distribution in the block is linear.
3. The horizontal stress is constant between roof and upper boundary.
4. Every layer has a constant thickness.
5. Every layer can be regarded as a fixed ended beam.

6a. All layers have the same thickness (i.e. the same stiffness).

Most predictions are slightly too high, because the ends of the beams are not completely fixed, so the excavation is somewhat smaller than 2l. Especially the last assumption limits the use of this analytical solution, therefore it is better to replace it by another assumption:

6b. All layers have the same elastic deformation.

This assumption results in a new condition, to wit, failure in the thickest layer

$$\frac{\sigma_v}{\sigma_h} = \frac{4}{3} \cdot \frac{(dh)^*}{l^2} \cdot \frac{D}{D-H} - \frac{d\gamma}{\sigma_h} \quad \text{with:} \quad (dh)^* = \frac{1}{h_{\max}} \sum_{i=1}^{n} \left(h_i^3\right) \tag{10}$$

In most cases $(dh)^* \approx dh$, like the simulation shown in figure 5 with two different layers

$$(dh)^* = \frac{6 \cdot 0.6^3 + 5 \cdot 1.4^3}{1.4} = 10.73 \approx 11 \tag{11}$$

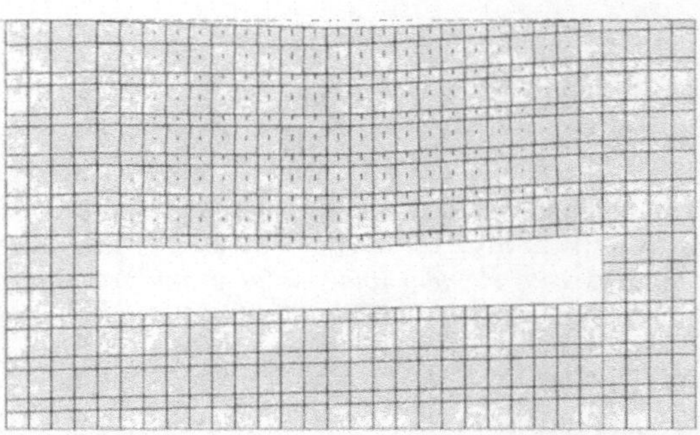

Figure 5. Beginning of failure in jointed rock with different layer thicknesses

With an excavation height H = 3.4 cm, the roof was expected to fail at σ_v = 108 kPa, but it collapsed already at σ_v = 87 kPa. This difference in vertical failure load can mainly be explained by assumption 5). The moments in the layers showed that the average beam length was almost $2l$ = 26 cm, which would have predicted a much lower and more accurate vertical failure load: σ_v = 78 kPa. It can be proved that the angle α between the layers and the horizontal is not so important, as long as tan $(\alpha) \ll \sigma_v / \sigma_h$.

It is shown that DEM can be helpful for obtaining and testing analytical predictions of the excavation problem. DEM can, of course, also be used for more complex problems, such as calculating the minimum required friction between jointed rocks to avoid a collapse of a tunnel. For example, the discrete model predicts that the friction in figure 6 has to be quite low to cause shear failure ($\tan(\phi) \leq 0.18$).

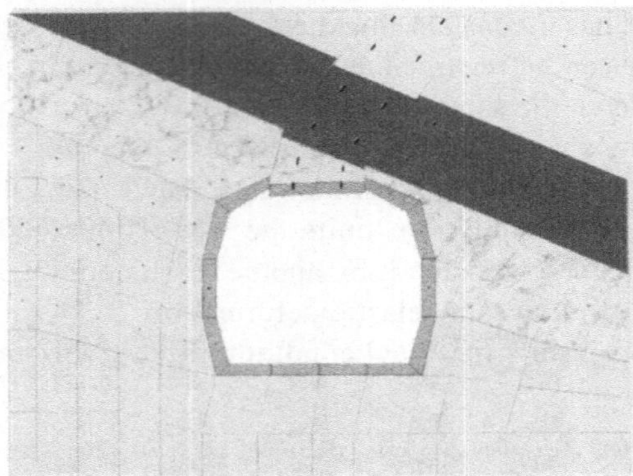

Figure 6. Shear failure in jointed rock

Acknowledgement

The author wishes to thank S. J. Mackenzie and J. T. Shepherd for their experimental results.

References

1. Baars, van S.: *Discrete element modelling of granular materials*. Communications on Hydraulic and Geotechnical Engineering, Report No. 95-4, University of Technology Delft, ISSN 0169-6548 (1995).
2. Baars, van S.: *Discrete element analysis of granular materials*, Dissertation, University of Technology Delft, ISBN 90-9009494-6 (1996).
3. Baars, van S.: Discrete element modelling of granular materials. *Heron 41(2)*, ISSN 0046-7316 (1996).
4. Cundall, P.A. and Strack, O.D.L.: A discrete numerical model for granular assemblies. *Géotechnique*, Vol. 29 (1979) pp. 47-65.
5. Mackenzie, S.J.: *Experimental investigation of linear arching in underground excavations*. Student report, University of Sydney, Civil Engineering appendix D, test 9a-9b (1996).
6. Shepherd, J.T.: *Behaviour of underground openings in jointed rock*. Student report, University of Sydney, Civil Engineering appendix B1, test 17a (1996).

The Principle of Virtual Work for Groundwater Flow

KLAAS-JAN BAKKER
TU Delft, Delft, the Netherlands
Ministry of Transport, Public Works and Water management

During his lectures on groundwater flow Verruijt used to begin with the fundamentals of a practical theory. for his lectures on groundwater flow, a number of lectures in sequence, he started writing down the fundamental equations; continuity of flow, Darcy's law, etc. After introducing boundary conditions problem and introducing simplifications, he explained, in detail, a specific flow problem, quite often with an analytical solution. It was as if he wanted to be sure that we learned the fundamentals of our job first, before he trusted practical problems in our hands.

When discussing the numerical modelling of groundwater flow, i.e. Finite Element analysis, he mentioned, though only slightly, that there was a minor fundamental problem with the numerical formulation of groundwater flow. Where for stress analysis one could rely on physical principles such as minimal potential energy for elastic problems, or the principle of virtual work for non-linear problems, such a physical principle for groundwater flow was not (yet) available. We had to rely on mathematical approaches, e.g. Galerkin's principle of variations. The analogy in the derivation, comparing the differential equations, made it trustworthy to rely on such principles, but on the other hand the lack of a physical principle was felt to be a missing.

Later on working with Rijkswaterstaat, when I was involved with the design of water defences this question on the fundamentals of groundwater flow, came back to me, when analysing the flow under revetment blocks. Looking into the literature, it could be found that among others Van derVeer [4] analysed this problem. He introduced what he calls the energy flux per unit of time: $e = \rho g q \phi$. He postulates that the loss in energy flux should be minimal. This would mean that the groundwater head distribution should be such that the total loss of energy should be a minimum. Working out the equations, the result is nearly the same as found using Galerkin, but the contribution of the boundary condition is not included. The limitation to Van.der. Veer's approach is that for fixed boundary conditions; i.e. groundwater flowing from a point with high groundwater head to another point with a lower groundwater head, whatever the further distribution the total energy loss, is constant. As the potential energy is the main energy component, the change in groundwater head visualizes the energy loss.

Groundwater flow is an energy non-conservative process. During flow continuously potential energy is transferred into heat. Therefore, one of the suggestions was that a more fundamental formulation would have to come from a thermo-dynamical, or more precisely for stationary flow, a thermo-mechanical formulation of the process. This implies the incorporation of temperature, and the temperature dependency of permeability. From a practical point of view, this does not lead to simple equations, and therefore, this approach was skipped.

Finally, a different approach is tried in analogy to the derivations for equilibrium of stresses in case of energy dissipation. The approach is based on Hamilton's principle of variations which is known for different physical problems, see Wylie [5] . It can be recognised with this, that the derived formulation for continuity of groundwater flow can be regarded as equivalent to the principle of virtual work, just as in the case of stress analysis.

Continuity of pore water

As a starting point the strong formulation of the storage equation is taken, equation (1), see Barends [1].

For numerical analysis this equation cannot be integrated directly to give a linear set of algebraic equations. Such a drawback is also know for the integration of the equations for equilibrium of stresses. In practicem for numerical analysis of elastic systems, the principle of minimum potential energy is preferred above direct integration of the equations for stress equilibrium. For systems dissipating energy the equation of virtual work is used, see Blaauwendraad [3].

For groundwater flow there is a similar problem, as it is also an energy dissipating; system a water particle moving along a streamline follows the negative gradient of the energy potential. The principle of virtual displacements as such is therefore not applicable. For practical purposes the mathematical principle of Galerkin is often applied, which gives the same equations as when applying Hamilton's principle, see the next section.

In addition, in the next section it will be shown that groundwater flow, both for steady and transient flow, essentially can be regarded and analysed according to the principle of virtual work, in the same way as the stress analysis.

Hamilton's principle for groundwater flow

Consider a groundwater particle moving in a field of groundwater flow, see Figure 1, and let $p(x,y)$ be the vector of groundwater pressure. According to the storage equation for small deformations, the continuity of groundwater flow in the body of a continuum V is determined by

$$\text{in } V \qquad \frac{\partial q_i}{\partial x_i} + \frac{\partial \varepsilon_v}{\partial t} + n\beta \cdot \left(\frac{\partial p}{\partial t} \right) = 0 \tag{1}$$

Where

q_i = groundwater discharge
x_i = axis in orthogonal co-ordinate system
ε_v = volume strain
n = porosity
β' = effective compressibility of the pore-water
p = pore-water pressure

For the boundary where a condition with respect to the discharge is demanded, denoted as S_1, see Fig 1, this means that:

$$on \quad S_1: \quad q_i \cdot n_i = \chi \tag{2}$$

where χ is the boundary discharge.

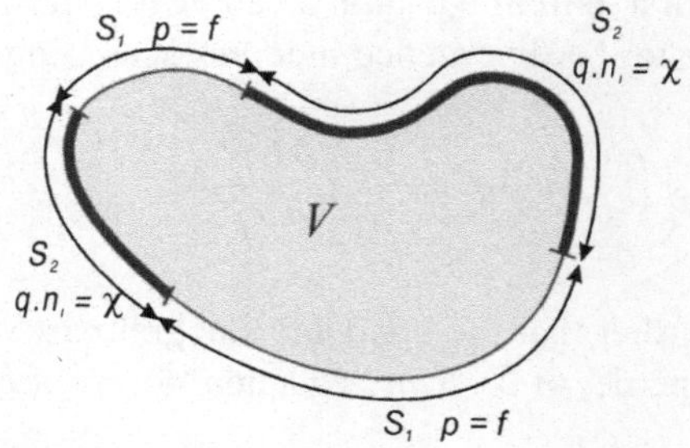

Figure 1.Two dimensional flow domain

For the boundary where a condition with respect to the groundwater head is imposed, denoted as S_2 a boundary condition for the pressure according to:

$$on \quad S_2: \quad p = f \tag{3}$$

is demanded.

The equations (1) to (3) are known as the strong formulation of the continuity equations for groundwater flow. This strong formulation however is not suited for a direct implementation in case of numerical analysis. In order to overcome this, the equations are integrated over the flow domain instead that equilibrium of flow is demanded for every point in the flow domain. The subsequent derived equations are known as the weak formulation. The derivation of the weak formulation can be found using Hamilton's principle, see Wylie [4].

In this derivations, the solution for these equations is sought such that an admissible pressure p is assumed with satisfies eqauation (3). Subsequently a perturbed groundwater head distribution $p + \delta p$ is assumed $p\ \delta p$ where δp is and arbitrary function of x, y, z and t, except for the fact that $\delta p = 0$

on S_2, i.e. the variation becomes zero for that part of the field where a pressure boundary condition is applied. If we subsequently form the scalar product of (the vector) δp with the left-hand side of equation 1 and integrate over the field, we obtain:

$$\int_V \left[\frac{\partial q_i}{\partial x_i} \delta p + \left(\frac{\partial \varepsilon_v}{\partial t} + n\beta' \left(\frac{\partial p}{\partial t} \right) \right) \delta p \right] dv = 0 \tag{4}$$

Applying integration by parts on the first term we obtain:

$$\int_V \frac{\partial q_i}{\partial x_i} \delta p \, dv = \int_V \frac{\partial \left(q_i \delta p \right)}{\partial x_i} dv - \int_V q_i \frac{\partial \delta p}{\partial x_i} dv \tag{5}$$

Subsequently the first term of equation 5 can be transformed into a boundary integral according to the divergence theorem according to:

$$\int_V \frac{\partial q_i}{\partial x_i} \delta p \, dv = \oint_S \left(q_i \, \delta p \right) \cdot n_i \, ds - \int_V q \frac{\partial \delta p}{\partial x_i} dv \tag{6}$$

If we recognise further that the boundary integral part vanishes for S_2, due to the assumed properties of δp, hence equation (6), equation 4 can be rewritten to give

$$\int_V \left[q_i \frac{\partial \delta p}{\partial x_i} - \left(\frac{\partial \varepsilon_v}{\partial t} + n\beta \left(\frac{\partial p}{\partial t} \right) \right) \delta p \right] dv = \oint_{S_1} (\chi \, \delta \, p) ds \tag{7}$$

Recognising that the product $\oint \chi \, p \, ds$ has the dimension [*m/sec* * *N* = *Nm/sec*], i.e. work per unit of time, the right hand side of equation (7) may be recognised as the virtual work (per unit of time) of the discharge on boundary S_1. In addition to that, the volume integral may be recognised as the virtual work of the groundwater flow in the soil mass.

If groundwater flow is being considered as such, i.e. if the storage part of equation (7) is of minor importance, this equation can be rewritten in terms of the groundwater head only, applying a simplified relation between Groundwater head and pressure according to;

$$p \approx (\Phi - z)\gamma_w$$

which gives

$$\int_V q_i \frac{\partial \delta \Phi}{\partial x_i} dv = \oint_{S_1} (\chi \, \delta \Phi) ds \tag{8}$$

It may be recognised from (8), that continuity in groundwater flow in a weak formulation can be modelled following the same line of reasoning as for structural mechanics, i.e. by applying the principle of virtual work.

Finite Element implementation

Quite often, equation (8) is further simplified, adopting Darcy's law, that the hydraulic discharge is proportional to the gradient in the groundwater head

$$q_i = K_{ij} \frac{\partial \Phi}{\partial x_i} \tag{9}$$

and using a simplified notation, for the groundwater-head gradient:

$$\psi_i = \frac{\partial \Phi}{\partial x_i} \tag{10}$$

using equations (9) and (10), equation (8) can be rewritten as:

$$\int_V \delta\psi_i K_{ij} \psi_j \, dv = \oint_{S_1} (\delta\Phi \chi) ds$$

or

$$\int_V \delta\psi K \psi \, dv = \oint_{S_1} (\delta\Phi \chi) ds \tag{11}$$

Next, a set of interpolation functions is adopted for a finite set of sub-domains (or finite elements) according to

$$\Phi(x,y) = N^T a \tag{12}$$

Where $\boldsymbol{\Phi}$ is the continuous deformation field, and a denotes the nodal groundwater-head at discrete points. The matrix N is a set of interpolation functions.

The relation between groundwater-head and groundwater-gradient is given by

$$\psi = L\Phi \quad 13)$$

Where here, L is the operator matrix

$$L = \begin{bmatrix} \frac{\partial}{\partial x} & 0 \\ 0 & \frac{\partial}{\partial y} \end{bmatrix} \tag{14}$$

Taking $B = LN$ and substituting equation (12) and equation (13) into equation (11) gives:

$$\delta a^T \left\{ \int_v B^T K\, B\, a \;\; dv - \oint_{S_2} \chi\, ds \right\} = 0 \tag{15}$$

As this equation must hold for any variation in the groundwater head, we obtain:

$$\int_v B^T K\, B\, a \;\; dv = \oint_{S_2} \chi\, ds \tag{16}$$

Concluding remarks

The implicit formulation for the continuity of groundwater flow such as given in equation (7) has a sound physical background, i.e. balance of the virtual work. Coming back on the lectures on groundwater flow, it was Verruijt who triggered the analysis of this problem.

The derived equations are not new. The thing new here is that these equations are discussed from a different point of view. The weak formulation for groundwater-flow, to which these equations for numerical analysis a rereferred to, is put in a physical context.

For stationary flow it has been shown that groundwater flow can be conceived in a similar framework as used in the analysis of equilibrium of stresses, i.e. with the principle of balance of virtual work.

References

1. Barends, F.B.J:.*Non linearity in Groundwater flow.* Doctoral Thesis, Delft University of Technology,1980.
2. Bakker, K.J.: Soil Retaining Structures. *Development of models for structural analysis,* Balkema, Rotterdam, 2000.
3. Blaauwendraad, J. and Kok A.W.M: *Finite Element method, part 1* (in Dutch), Agon Elsevier, 1973.
4. Van der Veer, P.: *Calculation methods for two-dimensional groundwater flow.* Doctoral Thesis, Delft University of Technology, 1978.
5. Wylie, C.R, and Barrett, L.C.: *Advanced Engineering Mathematics*, McGraw-Hill, 1986

Stability of a Vertical Cut

WOUT BROERE
A. Broere's Aannemingsmij BV, Amsterdam &
TUDelft, Delft, the Netherlands

When considering the local or micro-stability of a tunnel face, one often only considers the stability of soils with low or no cohesion in front of a slurry or air supported tunnel face. For the case of a slurry supported face in cohesionless material Müller-Kirchenbauer [2] presented a solution for the minimal stagnation gradient of the slurry required to ensure the stability of a small group of grains at the face. Combined with an estimate of the equivalent radius of the average flow channel between the grains, this solution can be used to estimate the minimal required yield strength of the slurry.

In cohesive soils one often disregards the local stability of the soil, as it is deemed secondary to the stability of the entire face. A problem could occur, however, in cases where a pressure gradient towards the face is present. This is easily conceivable in an earth-pressure balance shield with support pressures below the local hydrostatic pressure [1]. Unfortunately the general approach followed by Müller-Kirchenbauer does not lead to a straightforward solution in cohesive materials and a different approach is needed. Here we will consider the stability of a vertical cut, including the influence of a drag force in the direction of the cut.

Upper bound solution

The simple upper bound solution for the critical height of a vertical cut in a cohesive material is presented in many textbooks on soil mechanics, e.g. Verruijt [3]. For a vertical cut with height h assume a straight failure plane as sketched in figure 1. In this case the weight of the failure wedge is given by

$$W = \frac{1}{2} h^2 \gamma \tan \alpha \tag{1}$$

where γ is the volumetric weight of the soil and α is defined as in figure 1. From a simple force vector plot it can be found that

$$T = W \cos \alpha. \tag{2}$$

where T is the friction force, which is also equal to

$$T = \frac{h}{\cos \alpha} c \tag{3}$$

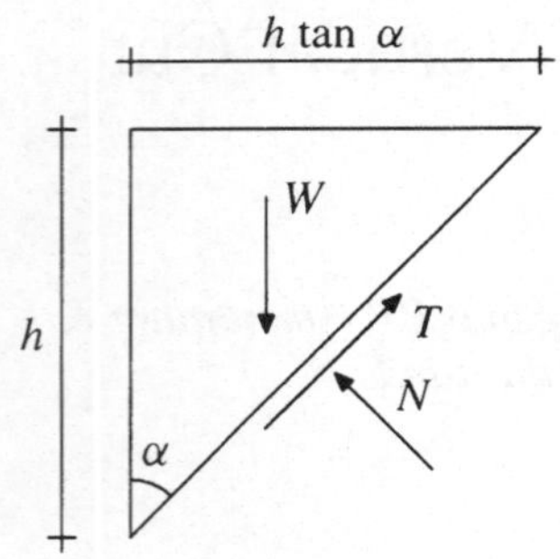

Figure 1. Definition of forces on a triangular failure body.

for a purely cohesive material with cohesion c. From the last two equations it can be easily derived that

$$h = \frac{4c}{\gamma}\frac{1}{\sin 2\alpha} \tag{4}$$

for a given slope α. Failure will occur for that value of α for which h is minimal, or $\alpha = 45°$. The critical height h_c is then

$$h_c = \frac{4c}{\gamma}. \tag{5}$$

As an aside it can be noted that in 1927 Fellenius found a lower upper bound solution using circular failure planes. His solution yields

$$h_c = \frac{3.83c}{\gamma}. \tag{6}$$

Influence of a drag force

These are well known results, but they do not include the influence of the drag force from groundwater flowing out of the cut. To investigate the influence of such a force a drag force I is added to the equilibrium diagram, see figure 2. Given a gradient i, constant over the wedge, there will be a pressure difference of $i\gamma_w z \tan\alpha$ over the wedge, where γ_w is the volumetric weight of water and z is a coordinate along the face, measured from the bottom of the wedge (see figure 2). Integration of this pressure difference over the height of the wedge results in the drag force

$$I = \frac{1}{2}h^2 i\gamma_w \tan\alpha. \tag{7}$$

Inclusion of this force in the force equilibrium diagram, see figure 3, shows

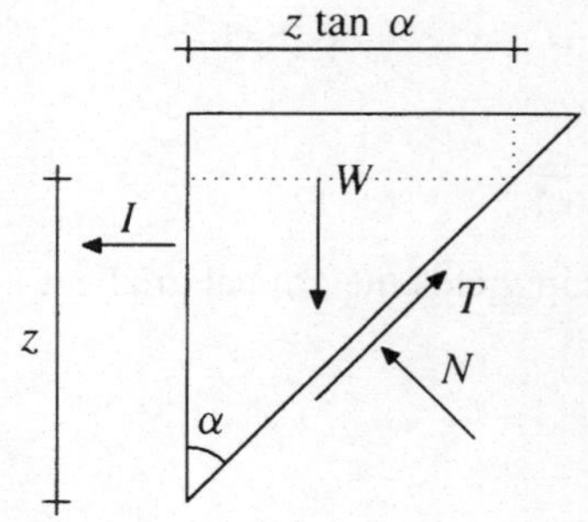

Figure 2. Definition of forces on a triangular failure body including a drag force.

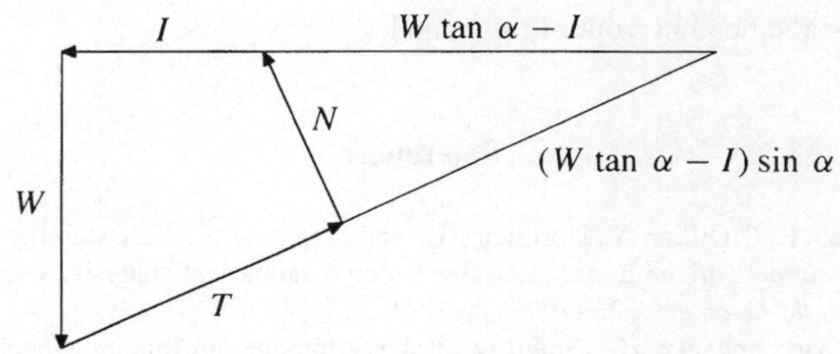

Figure 3. Force equilibrium including drag force.

that the friction force T can be found from

$$T = \frac{W}{\cos\alpha} - W\tan\alpha\sin\alpha + I\sin\alpha = W\cos\alpha + I\sin\alpha. \tag{8}$$

From this, using (1) and (7), it follows that

$$\frac{hc}{\cos\alpha} = \frac{1}{2}h^2\gamma\tan\alpha\cos\alpha + \frac{1}{2}h^2 i\gamma_w\tan\alpha\sin\alpha \tag{9}$$

so that

$$h = \frac{4c}{\gamma}\frac{1}{\sin 2\alpha + i\frac{\gamma_w}{\gamma}(1-\cos 2\alpha)} \tag{10}$$

which is the maximal stable height of the cut for a given angle α.

The critical height h_c can be found by minimizing (10) as a function of α. The critical value is obtained when

$$\alpha = \frac{1}{2}\operatorname{arccot}(-i\frac{\gamma_w}{\gamma}) \tag{11}$$

so that the critical height is given by

$$h_c = \frac{4c}{\gamma}\frac{1}{\sin\operatorname{arccot}(-i\frac{\gamma_w}{\gamma}) + i\frac{\gamma_w}{\gamma}\left(1-\cos\operatorname{arccot}(-i\frac{\gamma_w}{\gamma})\right)}. \tag{12}$$

Introducing $f = i\frac{\gamma_w}{\gamma}$ this can be simplified as

$$h_c = \frac{4c}{\gamma} \frac{1}{f + \sqrt{f^2 + 1}}. \tag{13}$$

Inverting this relationship yields the critical gradient i_c given a vertical cut of height h as

$$i_c = \frac{16c^2 - h^2\gamma^2}{8hc\gamma_w}. \tag{14}$$

Taking the height of the cut equal to the diameter of the tunnel or the height of the cohesive layer in question, this critical gradient can be used to obtain an estimate of the maximal allowable difference between the support pressure and the hydrostatic head in front of the tunnel.

References

1. de Buhan, P., Cuvillier, A., Dormieux, L., and Maghous, S.: Face stability of shallow circular tunnels driven under the water table: A numerical analysis. *Int. J. Numer. Anal. Meth. Geomech.*, 23 (1999) pp.79-95.
2. Müller-Kirchenbauer, H.: Stability of slurry trenches in inhomogeneous subsoil. In: N.N., editor, *9th International Conference on Soil Mechanics and Foundation Engineering*, 1977, pp.125-132.
3. Verruijt, A.: *Grondmechanica*. DUM, Delft, 1990.

Displacement Observations of Large Landslides and Their Interpretation

JAN D. NIEUWENHUIS
TUDelft, Delft, the Netherlands

If soil is observed to move along a slope AB the moving area is called a landslide (Fig.1). The definition is independent from measured displacements (*d*), computed velocities, the evolution of velocities with time or the detected presence of slide planes (*S1, S2*). A landslide therefore may either move slowly for indefinite periods, may accelerate or decelerate, or may end up in failure. The notion of movement is sufficient because movements are both a nuisance, and pose a threat. A nuisance, since movements create fissures disrupt drainage paths and provoke damage to private and public property, and a threat since slide planes tend to be smoothed and soil masses tend to get weakened by the movements. Hence, in principle, the resistance to failure decreases.

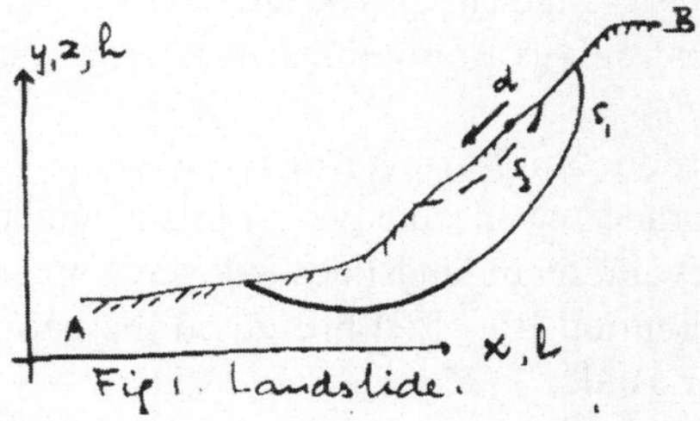

Figure 1

Landslides are mostly described by their displacements and velocities, very seldom by their accelerations. This seems odd, since the difference between driving force and soil resistance from stability analysis presents an excess force, the mass of the slide may roughly be estimated and the acceleration can be deduced from the velocity measurements. Newton's third law, force equals mass times acceleration, which holds for landslides as it does for every process in nature, looks easily applicable to landslide assessment. But if velocities are considered, 1 m/month for the Xintan landslide in China for instance, accelerations are of the order of 10^{-12} m/s^2. Even the impressive mass of large landslides, more than 10^{11} kg for La Clapière, can not explain more excess driving force than a fraction of a Newton! This embarrassing result is due to energy dissipation by rock breaking, heat production and tremendous viscous resistance and to much higher local accelerations through stick-slip movements not observed at the surface. Therefore, all estimates of deep landslide behaviour and unfortunately all predictions too are based upon correlations between soil resistance and estimates of slide plane geometries on the one hand, and deformation measurements on the other. Depending upon available data several ways are followed to arrive at an adequate description of landslide behaviour.

Landslides that failed have vanished as such and only so-called *back analysis* is feasible. Since thousands of landslides have been investigated in this way a lot of knowledge on their behaviour has been collected. Many mechanisms such as progressive failure, residual strength, creep effects, fissure flow amongst others have been first described in back analyses of landslides.

To protect human life and property large landslides are monitored. The shear size of a landslide and the consequent large variability of physical properties may strongly hinder the reliable estimation of slide plane geometry, soil strength and pore pressure values. The threatening landslide moreover may be unaccessable or moving to such an extent that soil coring or measurements are virtually impossible. In these cases analysis and monitoring forcibly are limited to *superficial displacement measurements*. Engineers and geologists from Japan have successfully applied the results of controlled experiments to predictions of the (creep) rupture life of landslides (Fukozomo [1]) through *1/v* (inversed velocity). Others undertook the analysis of the displacement (velocity) - time curve. The analyses run from harmonic analysis for clearly periodical curves to nonlinear dynamic system (NDS) analysis for less clear trends.

Without the introduction however of *relevant mechanisms* operating in the landslide, even crude and speculative, surprising and often most important trend breaks in the evolution of landslides will stay unpredicted In an amusing and sceptical contribution Rat ([2]) presented the capricious behaviour of the La Clapière landslide.

Displacement-time curve analysis

As soon as landslides move visibly, for the human eye this means faster than 0.3 mm/s involving accelerations 10^8 times larger than the Xintan landslide, failure almost always follows quickly. Public warning evidently must be issued earlier and at much lower rates. From many instrumented field experiments Japanese investigators deduced a relationship between acceleration and velocity v prior to failure.

$$\dot{v} = A v^{\alpha} \tag{1}$$

with A and a parameters to be determined from experiments. The integrated form of (1) :

$$\frac{1}{v} = \left[A(\alpha-1)\right]^{\frac{1}{\alpha-1}} \cdot \left(t_f - t\right)^{\frac{1}{\alpha-1}} \quad (2)$$

with t_f ultimate time for a failure warning and v and t some starting values, is used to extrapolate *1/v* to zero. The Japanese method works, provided a lot of experience is collected as was done by M. Saito and his coworkers at the Railway Technical Research Institute.
In figure 2 a time series of displacements at the surface of the large Xintan landslide (50 millions m^3 of soil and rock) is shown (Qin [3]).

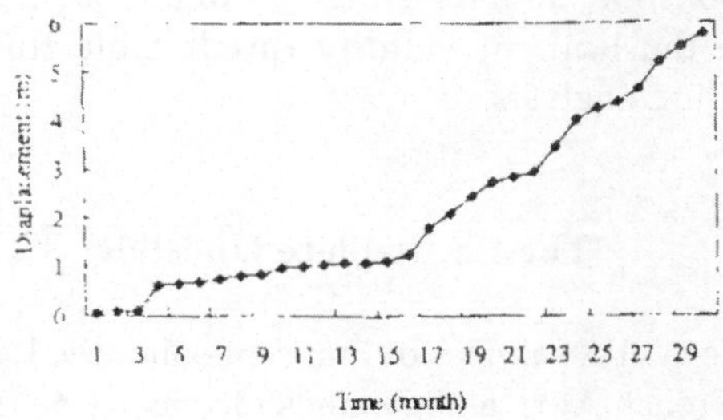

Figure 2. Displacements of the Xintan landslide in the Three Gorges, Yangtse river, China, July 1982-January 1985. (after Qin).

Even a mildly fluctuating time series such as the one from Xintan does not allow easy extrapolation of inverted velocities to zero; the values vary irregularly between 1 and 60 month/m. The last months, however, not reproduced by Qin, might have presented more adequate velocity values for the Fukozomo method. Anyhow, Qin et al. assume landslides to exhibit predictable and chaotic behaviour and they apply nonlinear dynamic system analysis (NDS) to the Xintan landslide. NDS was developed for fairly well-known complex systems described by a set of interacting components.

For landslides this would mean that the mutual interactions of lithology, soil strength, slope angle, hydrology, etc. are known and defined. If this is not the case, and it is not, NDS can be applied to a single time series by sampling components x_m in a so called *m*-dimensional phase space according to:

$$\begin{aligned} x_m(t_i) &= \left[x(t_i), x(t_i+\Delta t), \ldots, x(t_i+(m-1)\Delta t)\right] \\ x_m(t_j) &= \left[x(t_i), x(t_i+\Delta t), \ldots, x(t_i+(m-1)\Delta t)\right] \end{aligned} \quad (3)$$

with Δt, constant time interval and $/\ t_i - t_j / > \Delta t$.

If the time series considered exhibits regularities such as monotonic change or periodical behaviour then $x_m\ (t_i)$ from (3) shows regularities too due to the constant sampling interval Δt. Based upon the distance between a pair of points x_m, N^2 pairs in case of N points, the mathematics of *NDS* defines 2 parameters, *d* called the correlation dimension of the attractor and *h* the second order Renyi entropy. From d the degree of chaos or regularity of the time series is estimated and from *h* the predictable time scale *T* according to $T=1/h$ (Grassberger [4]).

The predictable time scale of the Xintan landslide, as computed by NDS appeared to be 5.2 months. The monitoring intervals of 15 days are quite acceptable and the suitable interval for deterministic predictions is relatively large. It must be coincidental that the time lapse between the end of the measurements (1 January 1985) and failure (12 June 1985) is 5.2 months. NDS-analysis looks like an interesting tool to define monitoring frequency and to estimate the limit of validity (predictable time scale) of more sophisticated failure analysis.

The La Clapière landslide

Since 1978 Public Works services of France monitor the La Clapière landslide in the Northern French Alps, a large landslide, nearly a kilometer long, more than a kilometer wide, deepest slide planes at 250 m of depth and 50 million m^3 of soil and rock mass (Follaci [5]). Due to its size and the inaccessability of its site the properties of the ground mass and the rocks along the slide planes are poorly known. Surface displacements have been measured on a quasi-continuous base since 1982 (Fig. 4).

Warnings for imminent danger were issued in 1983 (extrapolation) and 1987. A conceptual model of the slide including ground water level fluctuations correlated to the discharge of the local Tinée-river could explain the displacements up to 1987, though the maximum speed obtained, 80 mm/day, exceeded the expectation. And then during the autumn rains of 1987 the La Clapière slide slowed down quickly to a reassuring speed of 20 mm/day. The model, updated after every new surprise, could not explain nor include the renewed stabilization. Rat [2], the public officer responsable for La Clapière and for many more landslides in France, argued that the method followed by always improving the measurements burdened the prediction method with new pseudo-periodical fluctuations to be included without understanding the mechanisms involved due to lack of knowledge of the interior of the slide.

In the meantime the La Clapière slide has behaved tranquilly due to fortunate climatical conditions and Rat went happily into retirement. The abrupt stabilization was attributed initially to the progressive breaking up and fissuring of the ground mass above the slide planes due to the large displacements and the consequent larger permeability. Though Rat believed it, he argued that the increasing rains of the fall of 1987 would have penetrated easily in the slide's mass and would have caused a new acceleration. And that did not happen.

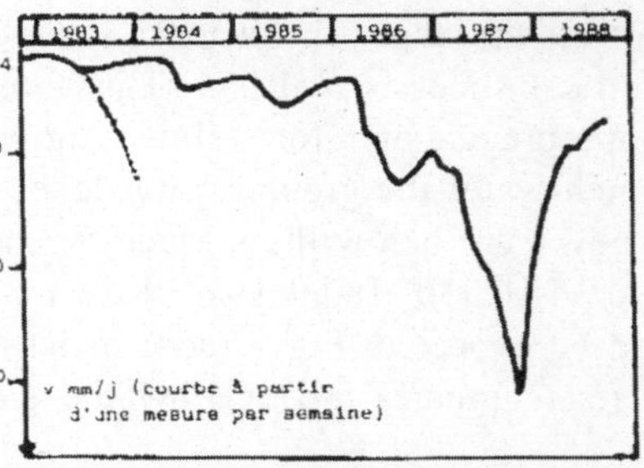

Figure 3. Displacements of the La Clapière landslide in France (after Rat)

Verruijt [6] drew renewed attention to the stabilizing effect of the downslope movement of slides along circular surfaces. To Bishop's method

$$F = \frac{\sum \dfrac{c+(\gamma h-p)\tan\phi}{\cos\alpha(1+\tan\alpha\tan\phi/F)}}{\sum \gamma h \sin\alpha} \tag{4}$$

with *c* cohesion, *f* angle of internal friction, ? *h* total weight of a vertical slice of unit width, p: pore pressure at the slide plane, *a*: slope angle of the embankment and *F* factor of safety, Verruijt added a new formula for a slide displaced over an angle *ß* (measured at the centre of the slip circle)

$$F = \frac{\sum \dfrac{c\cos(\alpha-\beta)+[\gamma h\cos\alpha - p\cos(\alpha-\beta)]\tan\phi}{\cos\alpha[\cos(\alpha-\beta)+\sin(\alpha-\beta)\tan\phi/F]}}{\sum \gamma h[\sin\alpha\cos\beta+(y_0-y_c)\sin\beta/R]} \tag{5}$$

with *ß*: angle of rotation due to displacements,
y_0: initial vertical coordinate of the bottom of a slice,
y_c: vertical coordinate of the centre of the slide circle,
R: radius of the circular slide plane.

The impressive increase of the factor of safety for even small values of *ß* inspired to consider the displacements of the La Clapière landslide rather from Verruijt's vieuw than through more sophisticated displacement analysis. The most dangerous slide plane of the landslide measures 855 m along the slope, slope angle 40°, and the deepest point reaches 250 m. The plane is roughly circular with a radius of 710 m.; it leaves the slope roughly horizontally: no Koppejan effects. Although the mean depth of the essentially 3D-slide is not more than 60 m the analysis is limited to the deepest circle. In view of the continuous displacements, since 1983 amounting to 40 m in 1987, the cohesive part c of the ground resistance is disregarded. Angle of internal friction ϕ estimated at 40°, wet specific mass of soils and rocks 2300

kg/m^3. The ground water level for the slow moving situation runs 20 m below the surface and parallel to the slope. The factor of safety according to Bishop's method (4) with vertical slices is F=1.034. This looks like an acceptable figure, of course not a precise one, for a slowly moving landslide.
If due to rain and melt water the ground water level would rise 10 m and consequently runs now 10m below the surface the factor of safety of the slide would drop to F=1.010. Indeed so close to 1.0 that accelerated displacements could be expected. For a more or less stable ground water level as computed for the early fall of 1987 these displacements would continue.

Up to 1987 the predictive analysis applied could be tuned to the measured slide displacements. Hence, to stay on the prudent side in the estimation of a possible effect of stabilisation due to displacements along a quasi-circular slide plane, only the last 8 m of the monitored 40 m of displacement, representing the displacements of the second half of 1987, are considered through Verruijt's analysis. If (5) is applied to the La Clapière slide while retaining the elevated ground water level 10 m below the surface and allowing 8 m of displacement which implies $\beta = 1°$, then a new factor of safety F=1.029 is found. This figure is 2.9% higher than the 1.01 according to Bishop's method and only a fraction smaller than the initial figure of 1.034. The stabilizing effect described by Verruijt is clearly strong enough to slow down the La Clapière slide in its 1987 condition. The soil masses deposited by the constantly collapsing toe of the slide are actually present at the site of the landslide. They assist the stability (Nieuwenhuis [7]). Deterministic models, even if globally and crudely applicable, may predict trend breaks; they should always be used! Verruijt's analysis is an example. A word of caution: the fortunate stabilization of the La Clapière slide, probably due to the displacements themselves, may easily turn into an unstable condition again due to the development of shallower planar slide planes!

References

1. Fukuzomo, T.: A new method for predicting the failure time of a slope. *Proc. 4th Int. Conf. and Fieldworkshop on Landslide* (1985), paper 10, Tokyo.
2. Rat, M.: Problèmes posés par la prévision de la rupture des movements de terrain. La Clapière, France. *Proc. 5th Int. Sympos. on Landslides*, Lausanne Vol 3 (1990), pp. 1503-1504
3. Qin, S.Q., Jiao, J.J., Wang, S.J.: The predictable time scale of landslides. *Bulletin Eng.Geol. and the Environment,* V59(4), (1986) pp. 307-312.
4. Grassberger, P., Procaccia, I.: Measuring the strangeness of strange attractors. Physica D9 (1983), pp. 189-208.
5. Follaci, J.P., Guardia, P., Ivaldi, J.P.:Le glissement de La Clapière (Alpes Maritimes, France) dans son cadre géodynamique. *Proc. 5th Int. Symp. on Landslides,* Lausanne, Vol 2(1988), pp.1323-1327.
6. Verruijt, A.: Computational Geomechanics, ch. 17, Kluwer Academic Publishers (1995), Dordrecht.
7. Nieuwenhuis J.D.: Experimental investigations into temporary blocking mechanisms observed in seasonal landslides. *Proc. 5th Int. Symp. on Landslides*, Lausanne, Vol. 1 (1988), pp.735-740.

Tension Pile Load Tests in a Glacially Overconsolidated Clay

FLORIS SCHOKKING
GeoConsult, Haarlem, The Netherlands

Various researchers have been working on the development of design methods for tension piles in overconsolidated clays, among which de Beer [1], for the Boom Clay in Belgium and Vesic [2] in the United States. Both used the concept of an adhesion value related to either a CPT friction value (de Beer) or the undrained cohesion as measured in the laboratory (Vesic).

In The Netherlands for the calculation of shaft resistance of piles in clays generally the q_c – method, as originally defined in the Dutch Standard for compression piles [3] and [4], is used, as well as a slipformula relating to the soil/pile friction angle, the coefficient of horizontal stress and an adhesion value. For this last value a parameter related to the undrainded cohesion can be used in the $\phi = 0$ case [4].

Research results of driven large diameter piles in the glacially overconsolidated Pot Clay, occurring in the northern part of The Netherlands and the adjacent part of the North Sea, are scarce. Fortunately, the results of two tension pile load tests, performed northeast of the City of Groningen could be interpreted by back calculation with the q_c – method and the slip formula. Promising results of values calculated with the slip formula for open ended piles in comparison with the measured shaft resistance were found, using generalised parameters from previous research in the same region [5], consisting of undrainded triaxial tests *f*-values and *K*-values with the Marchetti dilatometer. With more evidence from future tests this approach could prove reliable for the design of large-diameter, open ended piles in this specific overconsolidated clay.

Introduction

Some of the electricity masts of SEP's 380 kV connection from the City of Zwolle to the Eems Port (Province of Groningen) were founded on large diameter piles in glacially overconsolidated Pot Clay. GeoConsult worked together with the KEMA, on pile design. Four piles of which two were for a large part in the Pot Clay, were test loaded in tension during the spring and summer of 1995 to validate the design bearing capacities.

Geological and geotechnical setting

The geological sequence at the two sites of the pile tests consists of soft cohesive Holocene deposits underlain by a glacially overconsolidated and fissured clay, known as Pot Clay, in the region (Fig. 1).

The Pot Clay is deposited in a lacustro-glacial environment during the Elsterian glaciation in isolated occurrences of several square kilometer in area, up to 125 m in thickness and positioned on top of channel deposits in the Provinces of Groningen and Friesland and in the offshore area North of and West of The Netherlands. Extensive geotechnical research [5] was performed near Noordbergum, Marum and Beerta, of which the last two locations are positioned geographically some 30 km southwest and southeast of the sites of the pile load tests.

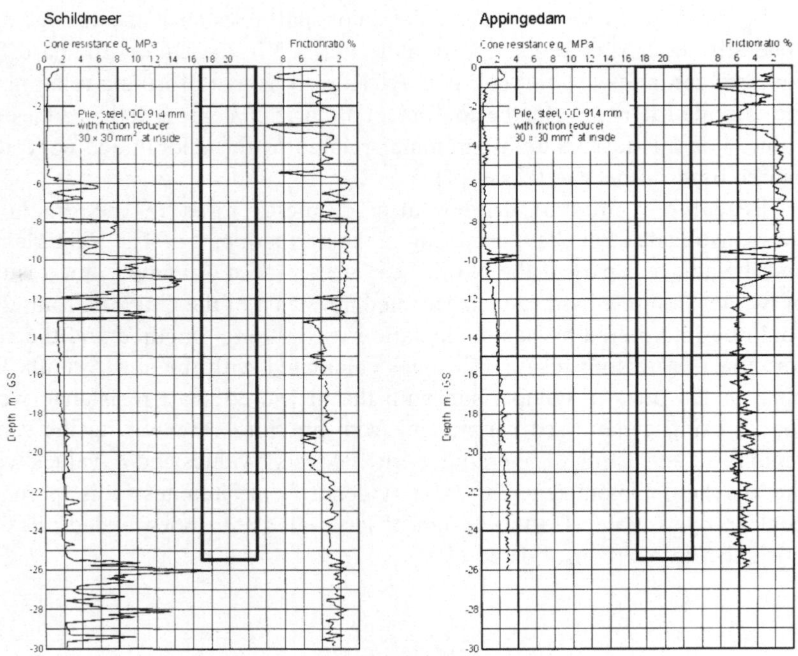

Figure 1. CPT with schematised tension pile; Schildmeer (A), Appingedam (B)

The Pot Clay has been overconsolidated by the ice load of the Saalian glaciation with maximum past overconsolidation stresses of σ_{vm}' of approx. 2100 kN/m^2.

At low confining stresses (< 400 kN/m^2) this clay shows no cohesion intercept and has an angle of internal friction of $\phi = 25^\circ$ to 35° as measured in consolidated undrained triaxial tests. At higher confining stresses the clay shows a cohesion intercept and a lower angle of internal friction, the material behaving more as an intact clay. The former behaviour is ascribed to the fissured character of the clay, which result in dilation along fissures comparable to the dilation of sands or of jointed rock.

The value of K measured in-situ in the Pot Clay at Marum with a Marchetti type dilatometer, and to similar depths as at the sites of the studied pile load tests, is approx. 1.5 to 1.2, and is decreasing in a downward direction [5].

Methods for calculation of shaft resistance

A back calculation of the maximum developed shaft resistance was performed using two different methods: the q_c- method, which is generally used for the calculation of the shaft resistance for compression piles [3] and [4] , and the slip formula [4].

q_c- Method

In this method the developed shaft resistance is related to the cone resistance q_c as measured in a CPT (1).

$$F_r = \pi . D . \int_0^L \alpha_s . q_{c;z} \; . dz \qquad (1)$$

The factor α_s for cohesive soil is related to the depth/diameter ratio *z/D*, and has been adapted recently [4]. As this method is primarily developed for piles loaded in compression the reduction in vertical stress during loading in tension is taken into account with a reduction factor of 0.8.

Slip formula

This method is based on the Coulomb failure criterion relating to vertical effective stress, σ_v' the coefficient of horizontal stress K, the friction angle between pile shaft and soil δ, and related to the angle of internal friction ϕ, and an adhesion value, comparable to the cohesion intercept of clays [2].

$$F_r = \pi . D . \int_0^L \left(\sigma'_{v,z} . \mathrm{K}.\tan \delta'_z \; .' + f_{adh,z} \right) dz \qquad (2)$$

For sands factors are given to relate δ/ϕ and K to installation method and soil density (Van Tol [6], Rikken [7]).

De Beer [1] gives for the overconsolidated Boom Clay, as occurring in Belgium, a relationship between the local friction f_s as measured with CPT, taking into account the scale effect of fissures, and for steel piles with diameters larger than 0.4 m, the conversion factor from f_s to f_{adh} is 0.3. This factor applies to piles with closed end or with a plug formed during installation.

Vesic [2]gives values adhesion factors ranging from 0.3 to 1 for values of f_{undr} > 150 kN/m^2, and based on the back calculation from pile tests, but without giving indications of influence of installation methods, which could explain the wide range of the values stated.

As for the Pot Clay the behavior in consolidated undrained triaxial tests is according to $\phi \neq 0$ for low confining stresses this concept has also been used for the back calculation of the the pile load tests, with $\phi = 30\,^{o}$. The value of K measured in-situ in the Pot Clay at Marum is ranges from 1.5 to 1.2 (Schokking [5]). For these values at Marum extrapolation through the region over similar distances as from Marum to the sites of the pile load tests appears to be feasible as similar values were measured in a site investigation in the city of Groningen (H.F. van der Sluis, pers. comm.) and in a research near the city of Bremen in northern Germany (Feeser [10]). This feature of regional similarity is a result of the horizontal stress parameters being related to sub-glacial deformation processes under the moving land ice cover during the Pleistocene, which are assumed to have been constant over large distances, as far as the properties developed in the Pot Clay are concerned.

Interpretation of pile load tests

At Schildmeer and Appingedam 914 mm diameter; steel piles, raking at 19° were tested up to failure. The piles, with an open bottom, were driven with no plugging, as a result of the use of a friction reducer of 30*30 mm placed in the inside of the pile. The tests was performed about one month after installation.

The loads at failure were 4450 and 4200 kN for Schildmeer and Appingedam, respectively. Without the shaft resistance from other layers, the pile's own weight, the weight of the soil inside the pile, the under pressure developed the shaft resistance for the Pot Clay layers are $F_{r;meas}$ = 2768 and 3241kN. The results of the back calculation for the two pile tests are given in Tables 1 and 2.

A comparison between the three calculation methods used indicates, that the slip formula with $\phi \neq 0$ gives results which for both tests approaches best the measured values. Both for the q_c-method and the slip formula with f_{adh} give the two tests rather deviating results.

The comparison between the q_c-method and the slip formula method with $\phi \neq 0$ comes down to the reliability of the α_s values for clays, which embody δ/ϕ and the horizontal stress, against the actual values of f and K as measured in the laboratory and in-situ. Moreover the q_c-method has the disadvantage that α_s values for clays do not relate to the pile type or an installation method. Where closed ended piles or piles with plug forming are used the disturbance of the soil around the pile can be severe, so that the given α_s values may not be applicable. This appeared to be the case in an earlier tension pile load test in Pot Clay on a closed ended pile, at Meeden-Diele, (J.F. Rikken, pers. comm.), which was not tested up to failure.

That for the slip formula with $\phi \neq 0$ the calculated values are closer to the measured results, is probably attributed to the installation of the piles, with only little disturbance. For displacement piles with closed bottom or in case of forming a soil plug during driving the slip formula is not applicable to the Pot Clay.

Table 1. Pile Test Schildmeer OD 914 mm $F_{r;meas} = 2768$ kN

m – GS	q_c-method				Slip formula method			
	$q_{c;\,aver}$ MPa	Z/D	α_s	$F_{r;calc}$	$f_{s;\,aver}$ MPa	$\varphi = 0$ $f_{adh} \rightarrow f_s$ $F_{r;calc}$	$\varphi = 25°$ $\delta/\varphi = 0.7$ K	$F_{r;calc}$
13.0 to 19.0	1.75	<20	0.025	603	0.05	861	1.4	1110
19.0 to 21.0	2.20	>20	0.025	379	0.11	947	1.3	601
21.0 to 25.5	2.00	>20	0.025	517	0.07	904	1.2	966
	F_r Pot Clay total			1499		2712		2677
	Factor $F_{r;meas}/F_{r;calc}$			1.87		1.02		1.03

Table 2. Pile Test Appingedam OD 914 mm $F_{r;meas} = 3241$ kN

m – GS	q_c-method				Slip formula method			
	$q_{c;\,aver}$ MPa	Z/D	α_s	$F_{r;calc}$	$f_{s;\,aver}$ MPa	$\varphi = 0$ $f_{adh} \rightarrow f_s$ $F_{r;calc}$	$\varphi = 30°$ $\delta/\varphi = 0.7$ K	$F_{r;calc}$
9.5 to 12.5	1.25	< 20	0.025	431	0.06	517	1.5	404
12.5 to 16.0	1.75	< 20	0.025	603	0.10	1004	1.4	615
16.0 to 20.0	2.25	> 20	0.025	775	0.13	1492	1.3	867
20.0 to 25.5	2.75	> 20	0.025	947	0.13	2052	1.2	1446
	F_r Pot Clay total			2756		5065		3332
	Factor $F_{r;meas}/F_{r;calc}$			1.18		0.64		0.97

Conclusions and discussion

The results of two tension tests on large diameter, open ended, piles founded partly in the glacially overconsolidated Pot Clay, indicate that the developed shaft resistance in the clay is well predicted with the slip formula using unconsolidated undrained triaxial test results and values for the coefficient of horizontal stress measured with the Marchetti dilatometer, both from other locations in the region.

The use of open ended, large diameter piles appears essential in the Pot Clay to prevent disturbance to occur along the shaft during driving.

Although the use of an average measured value of ϕ is sufficient, the angle δ, can be determined more accurately, with an adaptation to a lower or higher value based on a CPT friction ratio. Coefficients of horizontal stress appear consistent throughout the region.

Acknowledgements

The author thanks SEP and KEMA for the permission to use the results of the pile load tests for this paper. J.F. Rikken, previously with KEMA, is thanked for the fruitful co-operation during the design work for the piles and for the discussions afterwards.

References

1. Beer, de E.: Sonderingen en funderingen op palen in klei. *Fugro Sondeer Symposium* Utrecht, 1977.
2. Vesic, A.: *A study of bearing capacity of deep foundations*. Final report, School of Civil Engineering, Georgia Institute of Technology, Atlanta, Georgia, 1976.
3. NNI.: Geotechniek; Berekeningsmethode voor funderingen op palen Drukpalen NEN 6743 Delft, 1991.
4. CUR: 2002-4 *Ontwerpregels voor trekpalen.* Gouda, 2001.
5. Schokking, F.: *Anisotropic geotechnical properties of a glacially overconsolidated and fissured clay.* Doctoral Thesis Technical University Delft, 1998.
6. Tol, van A.F.: *Foundation Engineering G80*, Technical University Delft, 1993.
7. Rikken, J.F.: *Interne notitie* CUR C 98, Gouda, 1998.
8. Carpentier, R.: *Studie van het grensdraagvermogen van palen geheid in stijve en gescheurde kleilagen. Bepaling van het schaaleffect.* Proefschrift van promotie aan de Rijksuniversiteit te Gent, 1976.
9. Schokking, F.: *Anisotropic strength behaviour of fissured overconsolidated clay in relation to Saalian glacial directions. Engineering Geology,* 49 (1998), pp.31-51.
10. Feeser,V.: *Geomechanisches Konzept zur Gefügeentwicklung und Spannungsgeschichte glazial überprägter Tone.* Diss. TU-Braundschweig, 1986.

Optimal Methods and Parameters for HighSpeedRaillink/A16

SJOERT E.J. SPIERENBURG
Royal Boskalis Westminster & BC HSL Brabant Noord

At present a number of large railway and motorway projects are under construction in the Netherlands among which the High-Speed-Railway Link (HSL) Amsterdam – Paris combined with the upgrading of motorway A16. In most cases these projects are realized within a Design/Construct and Maintenance contract. The contract states the functional requirements in terms of maximum settlements after construction and dates for the realization.

The track of the HSL and non-level infrastructure crossings with high embankments are projected over areas with soft sediments. There, the settlement requirements have a major impact on design, total costs and risk exposure.

Total settlements can be divided in primary and secondary compression. In areas with compressible soils the rate of primary compression has to be progressed in order to meet the requirements. As a result the functional requirements relate mainly to the secondary settlements. Time constraints and the derived construction rate are strongly related to the strength during construction. The consolidation coefficient determines the rate of primary compression.

In order to make a cost effective design with an acceptable risk level there is a demand for optimal and reliable methods of calculation. The geotechnical design code of the Contractors Combination HSL Brabant Noord is based on an integrated procedure for the determination of the design methods, design values of parameters and the interpretation of field data.

Review of literature

Generally, compression is calculated with the international C_α method, see Terzaghi [1]. The strain is defined by

$$\varepsilon = CR \log [(\sigma + d\sigma)/\sigma] + C_\alpha \log t/t_0 \qquad (1)$$

Primary compression is determined by the compressibility index CR. Secondary compression is determined by C_α.

Traditionally in the Netherlands the Buisman-Koppejan method is used, Koppejan [2]. Koppejan combined Terzaghi's principle of compression with Buisman's formula for secondary compression [3]. The difference with the

C_α method regards the calculation of secondary compression. In the Buisman-Koppejan method the logarithmic stress dependency also applies to the secondary compression whereas in the C_α method this term applies to the primary compression only.

Primary compression is constraint by consolidation. For the coupled process of consolidation and secondary compression numerical solutions have been developed by Suklje [4], Garlanger [5] and Den Haan [6].
Slope stability is following the Dutch codes evaluated with a drained analysis. The in-situ stress distribution is hydrostatic from the ground level. Shear strength is derived from triaxial tests at a limited strain level (2 %).

An alternative is a so-called undrained method, see Leroueil [7]. The pre-construction in-situ strength is related to the natural overconsolidated stress state. Pre-consolidation is the result of secondary compression, see Bjerrum [8].

Parameter study

For the HSL Brabant Noord cluster of the High-Speed-Railway Link a large number of oedometer and triaxial tests have been performed. For the interpretation of the results a database was developed. With the database parameter analyses have been made for spatial variation of properties, relation of index properties, the relation of mechanical properties and design values of parameters are determined with a statistical principle.

Index properties

The variation of moisture content over the region of peat with low bulk density and clay with higher bulk density follows from the relation of moisture content with bulk density, see Figure 1.

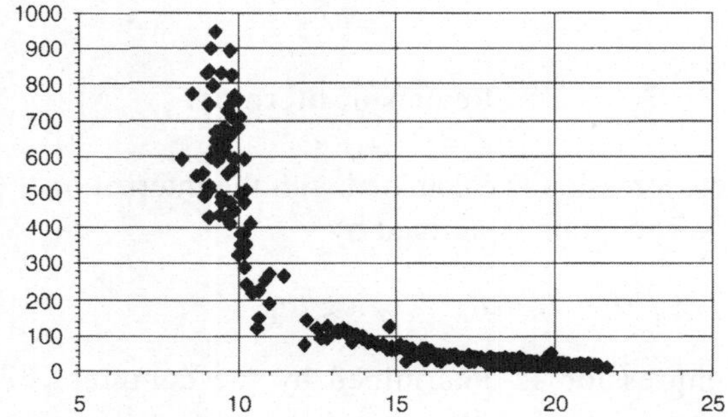

Figure 1. Moisture content [%] as a function of bulk density [kN/m^3]

Compressibility

Figure 2 *left* shows the relation of the compressibility ratio *CR* and bulk density. The results correspond to values found in literature, see [1]. In Figure 2 *right* the ratio of secondary and primary compressibility C_α/CR is given as a function of the bulk density. According to literature this ratio varies from 0.06 for peat (bulk density 10 kN/m^3) to 0.02-0.05 for clay, see [1] and [7]. C_α increases approximately linear with the applied stress. This conforms to the results in literature see Buisman [3]. It can be concluded that the properties of these materials conform to the experience in literature.

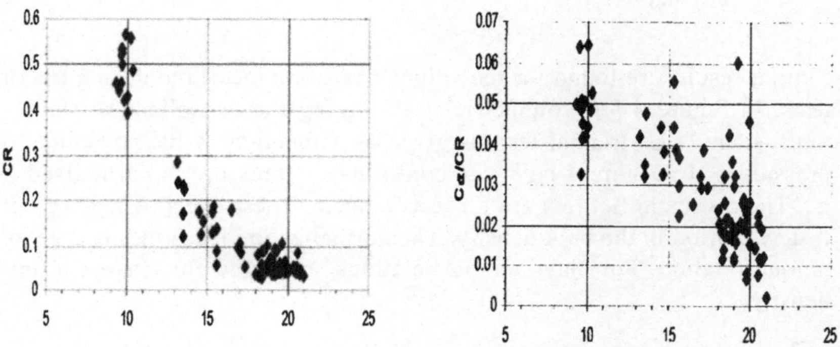

Figure 2. *Left*: Compressibility Ratio as a function of bulk density, *Right*: C_α/CR as a function of bulk density [kN/m^3]

Consolidation coefficient

The consolidation coefficient follows from the oedometer test either by the method of Taylor or Casagrande. Taylor's method leads to a higher value than the method of Casagrande, which is also generally found. The results of Taylor's method fit reasonably well to a value based on correlation with the liquid limit. Therefore the results of Taylor's method are used in the calculations.

Effect of stress history

Triaxial tests are interpreted in *p-q* space. However for natural clay deposits the pre-consolidation pressure σ_p determines the in-situ strength. According to literature the results of triaxial tests on natural clays should be normalized using the pre-consolidation pressure σ_p, see [7]. Using the database the results of triaxial tests were combined with the nearest oedometer test from the same geological unit in a boring.

In Figure 3 *left* the p,q points of tests on clay and peat of different depths are given and in Figure 3 *right* with normalization p/σ_p, q/σ_p. It is clear that the normalization leads to a reduction of the variation in q/σ_p and a more unique relation. Using the normalization the effect of the stress history is eliminated.

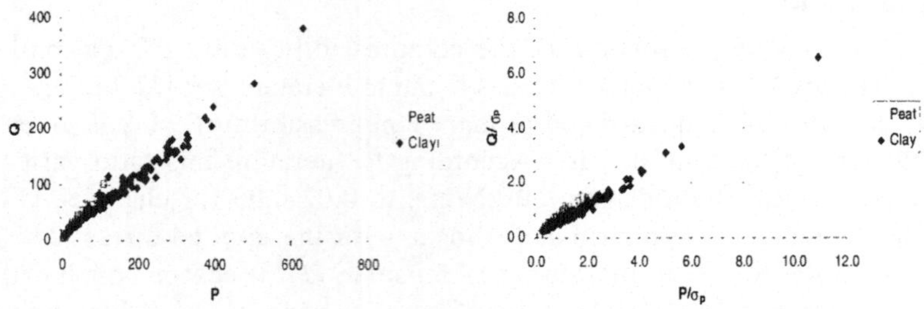

Figure 3. *Left*: Original p,q for peat (red) and clay (blue), *Right*: Effect of normalization p/σ_p and q/σ_p

Similar results are found for the volumetric strain measured during triaxial tests. In Figure 4 *left* volumetric strain during the consolidation stage of multi-stage (CU) triaxial tests is given as a function of the consolidation pressure p. In Figure 4 *right* the consolidation pressure is normalized by σ_p. The results in figure 4 are for tests on peat. These specimens have only little variation in the bulk density. The normalization leads in this case to a unique relation. For clays unique relations are found for classes in bulk density.

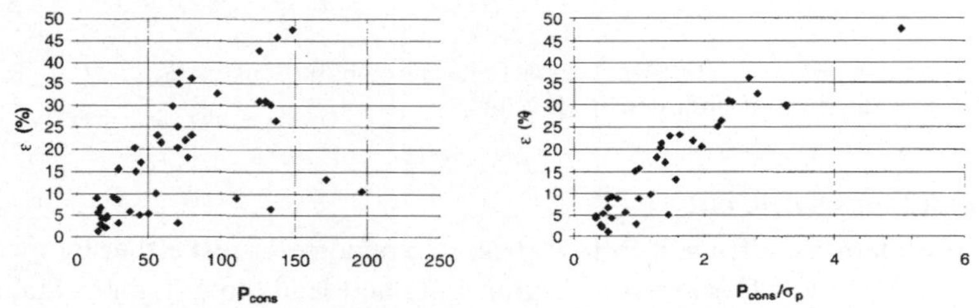

Figure 4. *Left*: Triaxial tests on peat, volumetric strain ε during consolidation phase as a function of p, *Right*: Effect of normalization p/σ_p

Effect of strain level on shear strength

The results of CU triaxial tests include the undrained shear strength at failure, defined by maximum q or maximum q/p and the drained parameters c' and ϕ', which vary with the strain level (2%, 5% or peak strain). For soft sediments such as peat the shear strength increases significantly with the strain level. This is shown by the calculated undrained shear strength using

$$c_u = c' \cos \phi' + \tfrac{1}{2}(\sigma_v + \sigma_h) \sin \phi' \quad (2)$$

In Figure 5 the calculated undrained shear strength is given for different strain level. The undrained strength increases by a factor 2 comparing results at 2 and 5 % strain. For the clay specimens the increased is much less as in most cases the peak strain is near 2 %.

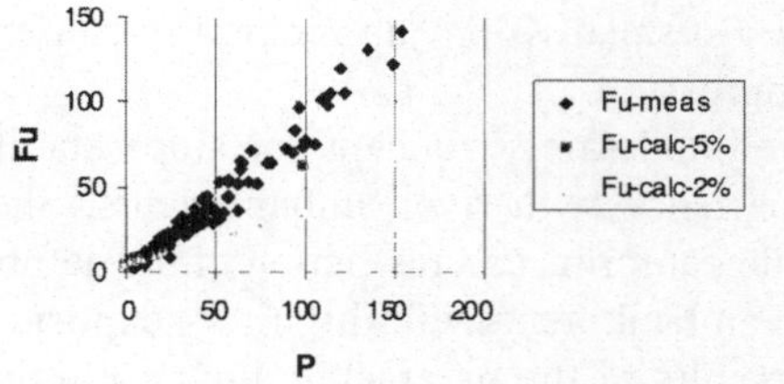

Figure 5. Peat: measured undrained shear strength *Fu* (blue) and calculated undrained shear strength from drained parameters at 2% (green) and 5 % (red) strain

Interpretation of field data

Based on cone penetration tests in and near existing embankments the in-situ compression could be determined. With the original thickness of the layers and the height of the embankment the compression was analyzed using average parameters for each layer from the database. Given the uncertainty due to spatial variability the results of the C_α method fit the compression in the field.

During construction settlements are measured at the ground level. Figure 6 shows the time settlement curve of an embankment. A back analysis is made with the C_α method. It can be concluded that the postdicted compression fits the data. The secondary settlements follow the C_α line.

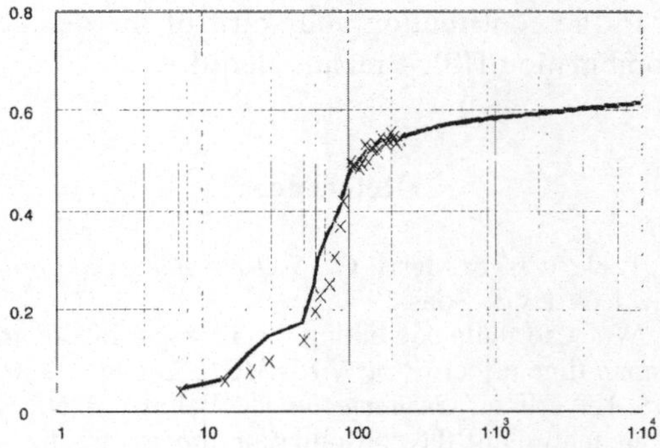

Figure 6. Measured time settlements (black) and postdiction with C_α method (red)

Discussion

The compression properties of the sediments conform to general data available in literature on the C_α/CR concept. In-situ compression conforms to the C_α method. Therefore, it can be concluded that the international method is an optimal and reliable method for the calculation of primary and secondary compression. The Buisman-Koppejan method is conservative with respect to secondary compression.

The design method and safety concept for slope stability in the Dutch is based on the experience with river embankments. Shear strength from a restricted strain rate conforms to a restriction of the deformations. This seems relevant for river embankments with high risk exposure.

Based on the results of the parameter study an optimization is possible for simple constructions. The effects of stress history, shear strength based on peak or near peak strain combined with an appropriate safety concept should be included.

For the design code of the HSL the required (total) stability factor was calibrated using international codes [9].

More attention should be given to the interpretation of soil tests e.g. the description of shear strength taking into account the effect of anisotropic yield function [7].

As shown in this paper an integrated procedure leads to optimization of the design. For this project a dataset was available. Projects in the future will benefit from regional databanks on a national scale. Such databanks should be developed in a cooperation of principals, consultants and contractors.

Acknowledgments

I am indebted to the contribution and spirit of the geotechnical team of Contractors Combination HSL Brabant Noord.

References

1. Terzaghi, K, Peck, R.B. & Mesri, G.: *Soil mechanics in engineering practice*. Third edition. J. Wiley & Sons 1996.
2. Koppejan, A.W.: A formula combining the Terzaghi load-compression relation and the Buisman time effect. *Proc II ICSMGE*, London 1936.
3. Buisman, A.S. Keverling.: *Grondmechnica*, Waltman, Delft, 1940.
4. Suklje, L.: The analysis of the consolidation process by the isotachen method. *Proc IV ICSMFE*, London, Vol.1 (1956), pp. 200-206.
5. Garlanger, J.E.: The consolidation of soils exhibiting creep under constant effective stress. *Geotechnique*, 22 (1972), pp. 71-78.
6. Haan, E.J. den: *Vertical compression of soils*. PhD thesis Delft University of Technology, 1994.
7. Leroueil, S, Magnan, J.P, & Tavenas, F.: *Embankments on soft clays*. English Edition, Translated from french by Wood, D.M., Ellis Horwood, 1990
8. Bjerrum, L.: Engineering geology of Norwegian normally-consolidated marine clays as related to settlements of buildings. *Geotechnique*, Vol. 17 (1967), pp. 81-118.
9. Calle. E.O.F. & Helden. G. van,: *Stability factor for HSL Brabant Noord*. Internal report, GeoDelft, Delft 2000.

Retaining Wall Stability Revisited

Erik O. Strack
TUDelft, Delft, the Netherlands

In 1906, Müller-Breslau [1] presented a generalization of Coulomb's famous solution [2] for the retaining force necessary to prevent the collapse of a retaining wall (see Figure 1). The active earth pressure coefficients computed from this

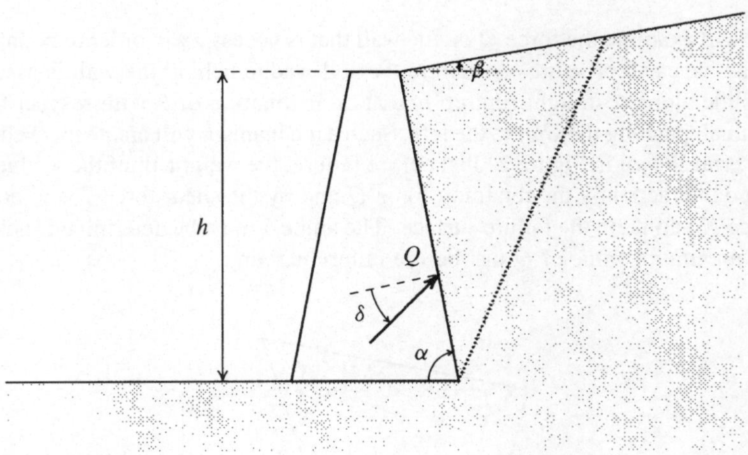

Figure 1. Retaining wall with possible rupture line (after Verruijt [3])

solution are still used in current engineering practice for retaining wall design. Although Müller-Breslau's solution can be found in many textbooks on soil mechanics [4, 5, 6, 7] (some of which quote Coulomb as the source of the formula [8, 9, 3]), the complete derivation has never been reported, to the best of the author's knowledge. Despite the fact that the solution is a straightforward application of the method used by Coulomb (from a mathematical standpoint), the calculation becomes quite involved and it is not easy to obtain the simple form of Müller-Breslau's final answer [10, 11]. A method of derivation which greatly simplifies the calculations, and which can be used in other maximization problems, is presented below in full detail.

Statement of the problem

We wish to obtain an equation for the minimum force necessary to retain a sloped soil mass with a rough and inclined retaining wall. The wall is drawn in Figure 1, along with a possible rupture line. It has a height h and is inclined at an angle of α, measured from the horizontal surface on which the wall rests. The surface of the backfill is inclined at an angle of β, also measured from horizontal. Due to the roughness of the wall, the retaining force Q exerted by the wall on the soil mass acts at an inclination of δ, measured from a line perpendicular to the wall. The soil is assumed to be cohesionless, and has an internal angle of friction equal to ϕ.

Determining an expression for the retaining force

We will examine the force Q on the wall that is necessary in order to counteract an active earth pressure. At failure the soil wedge behind the wall is assumed to slide along a straight rupture line at an inclination of θ with respect to the vertical, pushing the wall to the left. Such a mechanism will create the wedge of material shown in Figure 2. Just before failure, the weight W of the wedge will be counteracted by the retaining force Q and by the shear force T and normal force N acting on the failure surface. The angle θ must be determined such that the maximum value of Q just before failure is found.

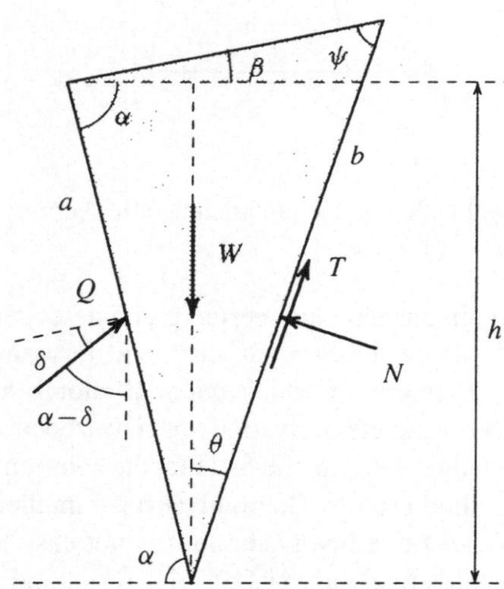

Figure 2. Forces on a wedge.

We begin by determining the values of the parameters a, b, and ψ shown in Figure 2 by using some of the basic trigonometric identities listed in Appendix A.

From the definition of the sine of the angle α we have

$$a = \frac{h}{\sin\alpha}. \tag{1}$$

From the sum of the angles of a triangle we know that

$$\psi = \tfrac{\pi}{2} - \theta - \beta. \tag{2}$$

From the law of sines (equation (A.1)) we have

$$\frac{\sin(\alpha+\beta)}{b} = \frac{\sin\psi}{a}. \tag{3}$$

It follows from (3), with (1) and (2), that

$$b = \frac{h\sin(\alpha+\beta)}{\sin\alpha\cos(\theta+\beta)}. \tag{4}$$

Next, we determine the forces acting on the wedge just before failure. Equilibrium of forces in the horizontal direction gives

$$Q\sin(\alpha-\delta) = N\cos\theta - T\sin\theta. \tag{5}$$

Equilibrium in the vertical direction gives

$$Q\cos(\alpha-\delta) = W - N\sin\theta - T\cos\theta. \tag{6}$$

At failure, the Mohr-Coulomb failure criterion states that

$$\tau = c + \sigma_n\tan\phi. \tag{7}$$

In our case, the cohesion $c = 0$. The shear stress τ and normal stress σ_n acting on the rupture plane are obtained by dividing the tangential and normal forces, T and N, respectively, by the length of the failure surface, b. After multiplying the resulting expression through by b this gives

$$T = N\tan\phi. \tag{8}$$

Substituting (8) in (5) and (6) and solving the resulting system of equations for Q gives

$$Q = \frac{W\cos(\theta+\phi)}{\cos(\theta+\phi-\alpha+\delta)}. \tag{9}$$

We can determine an expression for W through use of equation (A.2) in combination with (1) and (4). In fact,

$$\begin{aligned} W &= \gamma \times \text{area of triangle } abc \\ &= \frac{\gamma h^2\sin(\alpha+\beta)}{2\sin^2\alpha}\cdot\frac{\cos(\theta-\alpha)}{\cos(\theta+\beta)}. \end{aligned} \tag{10}$$

Substituting (10) in (9) finally results in

$$Q = \frac{\gamma h^2\sin(\alpha+\beta)}{2\sin^2\alpha}\cdot\frac{\cos(\theta-\alpha)}{\cos(\theta+\beta)}\cdot\frac{\cos(\theta+\phi)}{\cos(\theta+\phi-\alpha+\delta)}. \tag{11}$$

Note that this expression can be reduced to the expression obtained for a smooth vertical wall by setting $\alpha = \frac{\pi}{2}$ and $\beta = \delta = c = 0$ (compare Verruijt [3], equation 34.8).

Maximizing the expression for the retaining force

The expression for the retaining force (11) represents the force necessary for stability of the wall if the soil fails along a rupture line with an inclination of θ with respect to the vertical. In order to determine the minimum retaining force necessary to counteract failure along any inclination θ, we must determine the maximum of equation (11) with respect to θ. Setting the derivative of (11) equal to zero yields

$$\sin(\alpha+\beta)\cos(\theta+\phi)\cos(\theta+\phi-\alpha+\delta) = \sin(\alpha-\delta)\cos(\theta-\alpha)\cos(\theta+\beta). \quad (12)$$

The standard approach for solving the problem from this point on would be to find the limiting value of θ from (12), and then to substitute it in (11) in order to obtain an expression for the maximum value of Q. However, attempting such an approach leads to complicated equations which require a great deal of simplification in order to obtain the lucid form of Müller-Breslau's final result.

Simplification of the retaining force expression at its maximum

Instead of following the standard approach, we note that (12) can be used to simplify the expression for the maximum value of Q. From (12) we find that

$$\frac{\cos(\theta+\phi)}{\cos(\theta+\beta)} = \frac{\sin(\alpha-\delta)\cos(\theta-\alpha)}{\sin(\alpha+\beta)\cos(\theta+\phi-\alpha+\delta)}. \quad (13)$$

Substituting this result in (11) gives a convenient equation for Q when it attains its maximum (or minimum) value:

$$\begin{aligned} Q_{\text{max,min}} &= \frac{\gamma h^2 \sin(\alpha-\delta)}{2\sin^2\alpha}\left[\frac{\cos(\theta-\alpha)}{\cos(\theta+\phi-\alpha+\delta)}\right]^2 \\ &= \frac{\gamma h^2 \sin(\alpha-\delta)}{2\sin^2\alpha}\left[\frac{1}{\cos(\phi+\delta) - \sin(\phi+\delta)\tan(\theta-\alpha)}\right]^2. \end{aligned} \quad (14)$$

The number of times that θ appears in the expression for Q has now been reduced from four times in (11) to only once in (14). Note that this approach of back-substituting before computation of the actual limiting value can be utilized to solve other maximization problems as well.

Solution of the problem

It is clear from (14) that the problem can best be solved by finding a maximizing expression for $\tan(\theta - \alpha)$, rather than for θ. This can be accomplished by transforming (12) to the following form:

$$\sin(\alpha+\beta)\cos(\theta-\alpha+\alpha+\phi)\cos(\theta-\alpha+\phi+\delta) \\ -\sin(\alpha-\delta)\cos(\theta-\alpha)\cos(\theta-\alpha+\alpha+\beta)=0. \quad (15)$$

After expanding (15), dividing by $\cos^2(\theta-\alpha)$, and simplifying, we have

$$\sin(\alpha+\beta)\sin(\alpha+\phi)\sin(\phi+\delta)\tan^2(\theta-\alpha) \\ -\sin(\alpha+\beta)[\sin(2\phi+\alpha+\delta)-\sin(\alpha-\delta)]\tan(\theta-\alpha) \\ +\sin(\alpha+\beta)\cos(\alpha+\phi)\cos(\phi+\delta)-\sin(\alpha-\delta)\cos(\alpha+\beta)=0. \quad (16)$$

The coefficient of $\tan(\theta-\alpha)$ in this equation can be simplified further through expansion and use of the substitution $\cos[2(\phi+\delta)]-1=-2\sin^2(\phi+\delta)$, which corresponds to equation (A.6). In fact, we have

$$\sin(2\phi+\alpha+\delta)-\sin(\alpha-\delta)=2\sin(\phi+\delta)\cos(\alpha+\phi). \quad (17)$$

Using this substitution in (16) and solving the resulting quadratic expression for $\tan(\theta-\alpha)$ gives, after simplifying the expression under the square root sign,

$$\tan(\theta-\alpha)=\cot(\alpha+\phi)\pm\frac{\sqrt{\sin(\alpha+\beta)\sin(\phi+\delta)\sin(\alpha-\delta)\sin(\phi-\beta)}}{\sin(\alpha+\beta)\sin(\alpha+\phi)\sin(\phi+\delta)}. \quad (18)$$

We assume here (on the basis of Coulomb's simplified expression for the resultant force on a smooth vertical retaining wall) that if we take the negative root in (18), then the expression for Q in (14) will be maximized. Substituting (18) (with the negative root) in (14) gives, after simplifying,

$$Q_{\max}=\frac{\gamma h^2\sin(\alpha-\delta)}{2\sin^2\alpha} \\ \times\frac{\sin^2(\alpha+\beta)\sin^2(\alpha+\phi)}{\left[\sin(\alpha+\beta)\sin(\alpha-\delta)+\sqrt{\sin(\alpha+\beta)\sin(\phi+\delta)\sin(\alpha-\delta)\sin(\phi-\beta)}\right]^2}. \quad (19)$$

Dividing the numerator and the denominator by $\sin^2(\alpha+\beta)\sin^2(\alpha-\delta)$ and replacing Q with $\frac{1}{2}K_a\gamma h^2$, where K_a denotes the coefficient of active earth pressure, finally results in

$$K_a=\frac{\sin^2(\alpha+\phi)}{\sin^2\alpha\sin(\alpha-\delta)\left[1+\sqrt{\frac{\sin(\phi+\delta)\sin(\phi-\beta)}{\sin(\alpha+\beta)\sin(\alpha-\delta)}}\right]^2}. \quad (20)$$

This expression is equivalent to the Müller-Breslau formula [1]. An expression for the case of passive earth pressure (when the retaining force Q becomes so large that the soil fails and allows movement of the wall to the right) can be computed in a similar fashion. No attempt is made here to reproduce such a derivation. In fact, the solution for the passive case leads to dangerously high estimates of the actual force necessary to cause failure, since the rupture line will be far from straight [11].

References

1. Müller-Breslau, H.: *Erddruck auf Stützmauern.* Alfred Kröner, Stuttgart, 1906.
2. Heyman, J.: *Coulomb's Memoir on Statics, An Essay in the History of Civil Engineering.* Imperial College Press, London, 1997.
3. Verruijt, A.: *Grondmechanica.* 5th ed., VSSD, Delft, 1999.
4. Kézdi, A.: *Erddrucktheorien.* Springer-Verlag, 1962.
5. Smith, C.N.: *Elements of Soil Mechanics for Civil and Mining Engineers.* 5th Ed., Granada, 1982.
6. Craig, R.F.: *Soil Mechanics.* 4th Ed., Van Nostrand Reinhold, Berkshire, England, 1987.
7. Cheng, L.: and Evett, J.B. *Soils and Foundations.* 4th Ed., Prentice Hall, 1998.
8. Terzaghi, K.: *Theoretical Soil Mechanics.* John Wiley and Sons, Inc., New York, 1948.
9. Das, B.M.: *Principles of Geotechnical Engineering.* PWS Publishers, Boston, 1985.
10. Gudehus, G.: *BodenMechanik.* Ferdinand Enke Verlag, Stuttgart, 1981.
11. Azizi, F.: *Applied Analyses in Geotechnics.* E&FN Spon, 2000.

Appendix A : basic trigonometric identities

The following trigonometric identities, when relevant, are applicable to the triangle in the figure:

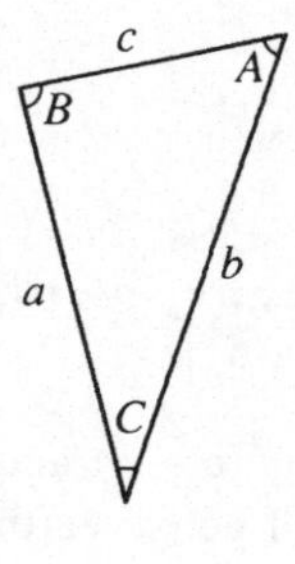

$$\frac{\sin A}{a} = \frac{\sin B}{b} = \frac{\sin C}{c}, \quad (A.1)$$

$$\text{Area of triangle} = \tfrac{1}{2}ab\sin C, \quad (A.2)$$

$$\sin(A \pm B) = \sin A\cos B \pm \sin B\cos A, \quad (A.3)$$

$$\cos(A \pm B) = \cos A\cos B \mp \sin A\sin B, \quad (A.4)$$

$$2\cos^2 A = 1 + \cos(2A), \quad (A.5)$$

$$2\sin^2 A = 1 - \cos(2A), \quad (A.6)$$